Anthony Reuben was the BBC's first head of statistics and now works on the corporation's fact-checking Reality Check brand, which he helped create. Anthony has 23 years' experience in journalism, and has been read by millions of readers of the BBC News website over the past 12 years. He has twice won the Royal Statistical Society's award for excellence in journalism and been shortlisted twice for the British Journalism Awards.

STATISTICAL

Ten Easy Ways to Avoid Being Misled By Numbers

Anthony Reuben

CONSTABLE

CONSTABLE

First published in Great Britain in 2019 by Constable
This paperback edition published in 2020

1 3 5 7 9 10 8 6 4 2

A CIP catalogue record for this book
is available from the British Library.

ISBN: 978-1-47213-025-9

Typeset in Stone Serif by Hewer Text UK Ltd, Edinburgh
Printed and bound in Great Britain by Clays Ltd, Elcograf S.p.A.

Papers used by Constable are from well-managed
forests and other responsible sources.

Constable
An imprint of
Little, Brown Book Group
Carmelite House
50 Victoria Embankment
London EC4Y 0DZ

An Hachette UK Company
www.hachette.co.uk

www.littlebrown.co.uk

For Susan, Isaac, Emily and Boaz

Contents

INTRODUCTION

The most powerful question

There's no getting away from statistics, however hard you try. Open a newspaper and see how long it takes you to find the first item based on numbers. It might be a poll about what people think of the US president or the latest figures showing what has happened to wages. Perhaps the country is experiencing record temperatures or the head of the NHS is arguing that the organisation needs more money. We use them every day.

It's not just in the news either. The company you work for may be revealing its gender pay gap or your child's school could be contacting you about its funding for the year. Your friends could be arguing about who is the greatest English batsman of all time or whether they should buy a petrol or diesel car. Perhaps you are trying to choose between different loans or just deciding whether to take an umbrella out with you today.

There are numbers everywhere you look and not all of them should be believed. That figure you have just seen on social media is very enticing – it feels like a proper fact and it supports your case in an argument you've been having with your friends. You've seen a claim that pay growth is at its worst level since the Napoleonic wars – is that true, and is it a fair comparison? Before you share a statistic, you really need to challenge it and look a bit further into where it comes from.

Unfortunately, lots of people do not have the confidence to challenge the numbers they read. Do you? Journalists, even some very good ones, get nervous around statistics and will not investigate them the way they would any other sort of evidence. Politicians, academics, people from all walks of life aren't comfortable with interpreting and understanding the way numbers are presented to them. You find researchers at their wits' end trying to prepare briefings for ministers or presenters to get statistical arguments across in a way that will be absorbed. We can be very critical when presented with an argument, but people who confidently question what they hear on the radio, or read in a newspaper, will happily accept the figures in a news story without challenging them, and move on. Normally cynical people will read the results of a survey and click the share button without a second thought about whether it is robust.

This is partly because it is considered acceptable to be hopeless with numbers in a way that it isn't with words. I have lost count of the number of times colleagues and friends have told me they can't cope with figures, but I very rarely hear anybody admit to being unable to spell or construct a sentence. We classify certain jobs as ones for the numerate (accountants, rocket scientists, actuaries) and decide that people in other jobs do not need to be able to cope with numbers at all.

In my experience, the mistakes made with numbers are usually made not in the numbers themselves but in the words used to describe them. And that's good news for you, because it means it's much easier to correct yourself – people tend to be much more comfortable with words than they are with numbers.

But a lack of confidence with numbers is a problem because there are many questionable ones out there and it's easy to be misled. Some of them are deliberately there to mislead you and artificially strengthen an argument; some of them are misleading by accident, because the person presenting them did not have the confidence or knowledge to check; and some are just thrown in to make a story sound more authoritative.

Without an understanding of data it's very hard to follow what's going on in your country. If you don't have at least a rough idea of what the population is or how many people are unemployed or how many people migrate to or leave the country each year, lots of political arguments become meaningless. If you have ever watched Prime Minister's Questions and wondered why the prime minister is claiming that crime is falling while the leader of the opposition claims that crime is rising, you'll be completely bemused unless you know that there are two completely different sets of crime statistics, one based on a survey and one based on crimes recorded by the police. When the leader of the opposition says that a public service is desperately short of cash and the prime minister says it is receiving record funding, you need to know that record funding is the norm, otherwise services struggle to cope with a growing population and inflation.

Statistics don't need to be scary. If you can add, subtract, multiply and divide then you already have most of the tools you need to challenge the numbers all around you. And in this book you'll find the other tools you need to deal with the things you hear that set off alarm bells in your head.

Take, for example, a 2018 report about plastic consumption in the UK. Following lots of justified concern about plastic waste, sparked by David Attenborough's *Blue Planet II* towards the end of 2017, the government launched a consultation about whether it should ban single-use plastic straws. There was research that demonstrated that we use 42 billion single-use plastic straws per year in the UK. There was all sorts of interesting background material about market research and dividing EU figures by levels of economic output for each country; the statistics would take ages to reproduce if you wanted to check them and it would be a pretty daunting thing to do. Still, we know that loads of straws get used in the UK, and this is certainly a big number, so why not just share it?

Here's why: if you just divide the number of straws by the population of the UK (just over 65 million) it turns out the report is suggesting that people in the UK consume an average of about 650 plastic straws each per year – that's almost two a day. My children were pretty enthusiastic consumers of plastic straws before we invested in a set of reusable silicone ones, but even they would have struggled to get through 650 each a year. That's not to say that the policy was wrong or we shouldn't be concerned about single-use plastics. The questionable stats were a distraction from the environmental message.

Without trying to dismantle the methodology behind the claim we have come to a pretty robust conclusion about whether we should believe it. And that is what this book is about.

This book will teach you in ten chapters how to get to the simple questions such as how many straws is it per person, which will allow you to challenge figures painlessly. The key to it is that before tackling data there is one question we should ask every time we are suspicious about the numbers. This is the greatest gift that I am giving to every reader of this book, even those who are just skimming through the introduction while standing around in a bookshop: it's the most powerful question in journalism.

It's not a question that you ask politicians or chief executives such as 'Are you going to resign?' That's sometimes a good question, but it's not as powerful as this question that you should ask yourself every time you watch, hear or read a news story.

So here it is – fanfare please.

Is this reasonably likely to be true?

This was a question I was taught by my father – a scientist – to use when checking my homework. You don't need to go back and carry out the calculation again. It's enough to look at the original question and estimate approximately what you think the answer should be so you can see if your own solution looks to be in the right ballpark.

This is the great insight when you're trying to challenge numbers. Many people are scared of numbers because they think it's their job to be right to three decimal places. Unless you're the person doing the original research and coming up with the figures, that's not true. When you're trying to decide whether to believe something or look into it further, you need to know that it's approximately in the right area.

What makes the most powerful question so useful is that the sort of figures you can use to work out if something is reasonably likely to be true do not have to be anything like as copper-bottomed as the numbers you would have to use if you were doing the research yourself. You can use any old thing you find on the internet, you can use stuff that folk have told you in the pub or vague ideas you have in your head. The idea is just to find out whether a figure seems suspicious and whether you should be looking further into it or seeking help from an expert.

Take the notorious *Daily Telegraph* headline in June 2010: 'Public pensions cost you £4,000 a year'. The story was that the cost of paying public sector pensions was going to rise over the next five years to £4000 a year per household in the UK.

How would you set about finding out whether that was reasonably likely to be true? If you wanted to find out whether it was precisely true then you would have to find out the total cost of public sector pensions and divide it by the number of households, but if you know both of those figures off the top of your head then you probably don't need to be reading this book.

A better approach would be to think about what would be a reasonable estimate of how much the average household earns in a year. You could even think about what your own household earns, unless you consider yourself to be fabulously rich or unrepresentatively poor.

There are various figures used for this but if you're using anything around about £27,000 then you're on the right track.

Now consider how much tax you are paying on that. You could work that out precisely, but on this sort of income, taking into account the amount you're allowed to earn without paying tax, income tax and national insurance together will come to around about 20 per cent. But it doesn't matter if you have guessed a bit more than that or a bit less. The point is that the figure you end up with is not going to be dramatically more than £4000 in tax. So if that much from each household is to be spent on public sector pensions, you need to ask who is going to pay for the NHS or schools or any of the other services that the government needs to fund.

The clear conclusion is that this figure is not reasonably likely to be true. And indeed, later in the day the *Telegraph* change the headline on its website to 'Public pensions to cost you £400 a year'.

With only a little bit of elementary maths – dividing stuff by other stuff and using percentages – you can easily work out if something is close to accurate.

There will be hardly any sums to do and no unreasonably tricky mathematical concepts. As long as you hear the alarm bells at the right time you can treat stories with due suspicion, and go and seek expert help if necessary. No longer will you need to be scared by stories about the growing risk of things or threats that will bankrupt the economy. Armed with the most powerful question, you will be free to go about your life without being misled by dubious statistics and bogus claims.

I stepped out of the front door the other morning and it was raining. I was about to pop back inside and put on a raincoat but then I looked at the weather app on my phone and it said there was a 0 per cent chance of rain. So I walked through the park in the rain and got wet. If only I had had the confidence to ask the most powerful question, I could have stayed dry.

It's not just about avoiding being misled or getting wet. You will also find joy in having the confidence to challenge statistics from colleagues and friends. In a competitive job market, if you're the one who questions whether the figures in a report sound likely, you will be on the way to the top. I describe myself as being a numerical pedant – now you can be the numerical pedant in your crowd. It's great fun and very satisfying.

I too find myself trying to cope in groups with massively more numerical expertise than I have. When I was the BBC's head of statistics I was invited to a one-day conference in Oxford to discuss whether there should be a global statistics authority. When I arrived there were 20 people around the table and it turned out I was the only one who wasn't the head of a national statistics office or didn't have a professorship, a knighthood or a Nobel Prize. I was probably just there to make up the numbers, although that is generally frowned upon in such circles. When it was time to go round the table for everyone to introduce themselves I noticed that the more important a person was, the fewer words they needed to use. Everybody there knew who 'Joe from Columbia' was (it was Joseph Stiglitz, the Nobel Prize-winning economist). I introduced myself as 'Anthony from the BBC' and added 'I'm here representing all the users.'

I've written this book for all the users as well, especially the involuntary ones. We are all users of statistics whether we like it or not, and everyone needs the confidence to challenge the numbers they hear. I am here to tell you that you already have most of the skills you need to question the numbers you see every day.

CHAPTER 1

Surveys

Guilty until proven innocent

You will never open a newspaper without finding at least one story based on a survey. I never open my email without finding one. They claim to tell me what UK workers think, or what business leaders think, or which animals are most popular in particular countries. Let me say up front: I believe that surveys must be considered guilty until proven innocent because some of the most dubious statistics you will ever see come from them. But there is a huge range of credibility for surveys. Some of them are unbelievably spurious, based on about seven people's views of a subject they are unlikely to be honest about. Others are the best-available figures, providing insight into difficult subjects.

One of the highest-profile official statistics in the UK is the monthly unemployment figures and they're based on a huge survey. The Bureau of Labor

Statistics in the USA conducts an even bigger survey. It's important to be able to spot which surveys you can believe and which you should reject. Whether people believe huge, sweeping statements about what the population thinks really matters.

My favourite dubious survey of all time started with the claim that 'a Saturday night in costs hosts up to £118.29 on average'. Notice how the writer has combined 'up to' with 'on average' to give a completely meaningless figure. Also notice the oddly specific figure, which I assure you is not justified by what is to come.

The figure is based on the idea that people (mainly women apparently) are inviting four guests over to watch *Strictly Come Dancing* or *The X Factor* and buying refreshments for them. The spending per person is £11.24 on alcohol, £10.92 on takeaways, £6.23 on snacks (that's a lot of chips and dips) and £6.32 on soft drinks. That spending on soft drinks looks particularly high – I reckon you should be able to get about eight litres of fizzy drinks for that, enough to quench even the strongest thirst, especially when you have already consumed at least eight cans of beer or a bottle or two of wine. But it's the cost of a takeaway that I love the most because the press release includes the methodology behind this one. It's based on 'desk research' that has involved finding out how much set menus for four cost at Chinese restaurants in Cardiff, London and Manchester, and Indian restaurants in Fife, Nottingham and Bournemouth.

But I've saved the best aspect of the research until last, which is that their survey has established that 55 per cent of women bought a new outfit to wear in front of the television, 'inspired by the glamorous judges on *X Factor* and outlandish outfits on *Strictly* – spending up to £100'. It's a triumphantly terrible piece of work.

Later in the chapter, I'm going to take you through some more of my favourite dubious surveys and in the next chapter I'm going to look at opinion polls around elections. First, I'm going to run through how you can tell when surveys aren't as robust as they might be.

Imagine you're trying to find out what people in your country think about cats. How would you go about finding out? The most accurate way would be to ask everybody in the country what they thought about cats. You need to get your questions about cats to millions of people and make sure they answer. That's how censuses work. Every ten years a questionnaire gets sent out to every household and they are legally required to complete it. So you could put a question about cats on the census. The trouble is, the census is an expensive project and it takes ages, so you won't get to find out what people think of cats for years. And you need to know this week, for some urgent, cat-related reason.

That's why people conduct surveys. The idea is that if you can get a smaller group that has the same features as the whole population, you can just ask the smaller group and then say that's what everyone in the country thinks. Doing that properly is jolly difficult, but people try to do it all the time. If you see the results of a survey and are worried you are being misled, there are five questions to ask:

- Where has the survey come from?
- What questions does the survey ask?
- How many people were asked?
- Were they the right people to ask?
- Is the organisation making reasonable claims based on the results?

Where has the survey come from?

There are two parts to this question: who conducted the research and who paid them to do it?

The answers to these questions should not disqualify the survey immediately, but they will put you on your guard and make you look more carefully at the answers to the other questions. It helps if the organisation conducting the survey is a member of a body such as the British Polling Council because it means that the way it conducted the research should be easier to find, but it does not mean the research is necessarily reliable. Also, there are large research organisations that do some sensible work that are not members of such a body so it's not a definitive guide, but it is a useful shortcut.

You should be more suspicious if the organisation that has paid for the research has an obvious interest in the outcome, so if the report saying that everyone loves cats comes from a manufacturer of kitty litter then it should put you on your guard, but it doesn't necessarily mean that the conclusions are nonsense.

I received a press release saying that 30 per cent of people would consider taking a holiday at a site affected by radiation. It was sent by a company that makes equipment that protects you from radiation. The figure might still have been true, and it may indeed be the case that holidays in Chernobyl are flying off the shelves. On the other hand, there were no details given of how they reached that figure of 30 per cent or indeed who had conducted the research. Do you want the news you hear to be influenced by this company? This is really one to avoid.

Another important thing to be aware of is that the organisation that conducted the research may just be using data from its own customers. So, for example, a health insurance company may be using the answers it has been given by its members when they sign up. This is a problem because it is extremely unlikely that they will be representative of the whole of the country – they're likely to be unrepresentatively wealthy for starters. I'll return to this when I talk about which people are asked later in the chapter, but beware research coming from organisations with access to lots of their own data.

What questions does the survey ask?

See if you can find what questions were being asked in the survey – if it's a reputable pollster they should be easy to find. If someone stopped you in the street and asked you those questions, would you be sure what they meant or are they ambiguous? Do you feel they might be trying to point you towards a particular answer? If you were asked, 'Do you just adore lovely, fluffy, beautiful, cute little kittens?' you might be more likely to respond with positive feline sentiments than if you were just asked, 'What do you think about cats?'

There was a scene in the classic BBC sitcom *Yes, Prime Minister* when the wizened permanent secretary Sir Humphrey asked young, inexperienced assistant Bernard a series of questions about whether he felt young people needed discipline and direction in their lives, leading up to the question of whether national service should be reintroduced. He then asked a series of questions about whether young people should be forced to take up arms against their will or given weapons and taught how to kill, leading up to the same question about conscription. It meant that Bernard agreed both that national service should and shouldn't be reintroduced, which Sir Humphrey described as 'the perfect balanced sample'.

Bernard was led by the earlier questions towards giving opposite answers to the final question. It's usually pretty obvious if a pollster is trying to lead respondents in a particular direction when you look at the questions asked.

Here's a real-life example: before the publication in 2012 of the Leveson Inquiry report into press standards in the UK, the polling organisation YouGov was commissioned to conduct different surveys by the *Sun* and the Media Standards Trust (MST). In the MST survey, 79 per cent of respondents agreed that 'there should be an independent body established by law, which deals with complaints and decides what sanctions there should be if journalists break agreed codes of conduct'. That seems to be at odds with the survey conducted on behalf of the *Sun*, in which only 24 per cent of respondents thought that the best way to regulate the press would be through 'a regulatory body set up through law by Parliament, with rules agreed by MPs'. This was the same polling organisation conducting surveys for different customers and getting fundamentally different answers to the same question. Except it wasn't quite the same question. YouGov's Peter Kellner explained that the difference lay in the way the question was framed, in particular the fact that the first question asked about an independent body established by law, while the second said it would be set up through law by Parliament, with rules agreed by MPs. While these two proposals are fundamentally the same thing, it turns out we don't like to think of MPs being involved with press regulation, so when they were mentioned it made respondents uneasy.

Another example is a survey that asked whether people were saving money in a pension to help pay for social care in retirement. Imagine how you would answer that question. You may be saving money in a pension, but might not have been thinking about social care when you started saving. Some people would have said 'yes' because they were saving, and so that money could help to fund their social care if they needed it. Others would say 'no' because they were not specifically saving to fund their social care. The question was not clear enough, so the results of the survey were completely unreliable.

Polling guru David Cowling sent me another example from the height of the financial crisis in October 2008, when the pollsters ComRes conducted two surveys of about 1000 people, asking what they thought about bailouts. One for the *Independent on Sunday* asked: 'Is it right that taxpayers' money should be used to bail out the banks?': 37 per cent said yes, 58 per cent said no and 5 per cent didn't know. In a survey the same month for the BBC's *Daily Politics*, the same pollsters asked whether people agreed with the statement: 'I support the government using taxpayers' money to stabilise the financial system': 50 per cent did, 41 per cent didn't and 9 per cent didn't know. Two polls in the same month conducted by the same organisation received very different answers to the same question. Presumably the difference was because one of them mentioned bailing out banks, which people were generally unenthusiastic about, while the other talked about stabilising the financial system, which respondents thought was a good idea.

You should also consider whether you think people are likely to answer the questions honestly. I was sent a survey of whether people smacked their children. I was assured that it was anonymised, but given that it is socially unacceptable to smack children, it means people are very unlikely to admit to it, even in an anonymous poll. Separately, I saw a survey of 16- to 18-year-olds, most of whom claimed that getting good qualifications was their top priority, they weren't very interested in sex and what they really liked was spending time with their families. I wonder if they thought they might be overheard by their parents.

The same problem with the likelihood of getting accurate responses is true when you conduct a survey of whether people have broken the law – the results will be unreliable.

How many people were asked?

If you're going to say something about a large number of people, it's no good just asking one or two of them. You would be amazed at the amount of press releases I see in which they have asked a few hundred people and are trying to use that to make claims about a whole country. Kellogg's based the claim that its new recipe for Coco Pops was loved by kids on asking 100 of them. And the claim that it was approved of by mums was based on asking 200 of them. The general rule is: if you're trying to say something about a population of 20,000 people or more, you need to have responses from at least 1000. Some of the smallest sample sizes you will see are on adverts for cosmetics, in which, for example, a company may say that 80 per cent of women agree that a particular lip gloss makes their lips look visibly plumper, and it's based on asking 43 women. I'm assured by the Advertising Standards Authority (ASA) that if the claim was about something more important such as whether a particular product will stop your baby having tummy ache, it would take more notice and demand proper evidence. The ASA doesn't care as much about how visibly plump your lips are. It's possible that a company could get away with making claims about lip plumpness without asking any women at all.

If you're trying to say something about a population smaller than 20,000, getting a representative sample becomes more difficult. If you want to know whether professional football managers like cats, for example, you would really have to ask all of them. That's because there are so few professional managers that the margin of error if you just asked a few of them would be enormous. Consider an extreme example – you are the conductor of a choir with ten members and you want to decide what combination of show tunes and Christmas carols to perform at your next concert. If you're going to base it on which is preferred by members of the choir, you couldn't get a reliable estimate of that by asking just half of your singers because there's no reason to believe that half would be representative – it may be that the other five would completely disagree. The only solution would be to ask everyone.

But there's another way round the problem of saying something about a small population. Everyone understands that it's difficult to get answers from all professional football managers, so they're not going to think any less of you if you have only got answers from half of them. 'We spoke to half of all professional football managers and found that most of them think cats are great,' is a perfectly reasonable, if slightly random, newsline. What you can't do with that data is claim that the majority of football managers think cats are great.

I used to get sent a regular survey of what chief financial officers (CFOs) from UK companies think. The company that calculated it spoke to about 100 CFOs every three months and sent out the results. That's fine – CFOs are not particularly easy people to survey, so if they had started with 'a survey of about 100 CFOs has found . . .' people would have understood what they were getting and that it might not be strictly representative. But, in fact, they always sent round press releases claiming to show what UK CFOs think as a whole. Talking to 100 of them isn't good enough to do that, even if you have the CFOs of a number of major companies, when there are many thousands of companies in the UK that could have a CFO. Compare that with the Ifo survey of business confidence in Germany, which hears from 7000 companies a month. Or the Tankan survey in Japan, which speaks to more than 10,000 companies a month.

You also need to beware surveys that have broken down their results further than the sample size will allow. For example, I was sent the results of a survey of

2000 people from five European countries, asking whether they ever pretended to be ill to get a day off work. First of all, this survey suffers from the same problems as the smacking one – would you be likely to admit to it? But 2000 is a perfectly reasonable sample size to say something general about the five European countries, or even to break it down into whether men or women are more likely to feign illness. What it couldn't do was then look at the skiving habits of French workers (who it claimed were the most likely to pretend to be ill) because it would only have spoken to about 400 of them. We'll return to the problems of unwarranted sick-day statistics in Chapter 8.

As the size of the population you're talking about increases, that 1000 minimum number of responses stays the same, but the next question gets harder.

Were they the right people to ask?

Speaking to the right people is remarkably difficult and if you haven't spoken to the right ones it doesn't matter how many people have been surveyed. Even if you interview a million readers of *Cats Monthly*, it's not going to tell you anything about what people in general (who do not, on the whole, subscribe to cat publications) think about cats.

Remember, you're trying to get a representative sample, so you need the relatively small group of people to whom you're talking to have the same qualities as the whole population about whom you're trying to say something. How do you go about doing that? If you stand on a busy high street and ask passers-by, the answers you get are going to be skewed towards people who live in the area, who might be richer than the country as a whole, or older, or more likely to be allergic to cats. And they will certainly be the ones with unrepresentatively large amounts of time on their hands because they have stopped and answered your questions.

There are two potential solutions to this problem and they each have their shortcomings.

The classic way of dealing with this is to get a completely random sample of people from within your population. A popular way of doing this is to get the phone book, randomly choose numbers from it and keep phoning them until enough people have responded, generally 1000. Nowadays, this skews the results towards older people who are more likely to have landlines and to answer them.

After the 2015 general election in the UK, Martin Boon from the pollsters ICM revealed that in order to get 2000 responses they were having to call about 30,000 random numbers, which raises the great danger that the people prepared to answer both the phone and the questions are not representative. When such a small proportion of people are answering the questions, you have to wonder if there is something unusual about them – perhaps they are older than the population as a whole or unrepresentatively interested in politics.

Many polling companies have started using mobile phone numbers as well, but only a relatively small proportion compared with landlines. The best practice way of getting a random sample is what the Office for National Statistics does with the Labour Force Survey, which is to choose households at random and then send people round to knock on doors, returning if nobody is at home. Once initial contact has been made, further interviews may be conducted over the phone. That's an expensive way of finding a sample, and most organisation's commissioning polls can't afford it.

The other way of getting what you hope will be a representative sample is to build up a panel of people who you know lots about. Then, if a client comes to you and says they want to do some research into what Belgian women who work full-time and like ice cream think about something, you can get in touch with them straight away, while making sure that the group you have is representative in other ways such as age and geographical spread across Belgium. You have to pay people a small amount to be on these panels, and the online access means it will be skewed towards younger people with internet access. There have been suggestions that the system provides incentives for people to lie about their characteristics to make themselves likely to be polled more often. And this gets to the nub of the problem – the people on your panel will be unrepresentatively interested in completing surveys. Also, it's unclear where you stop in identifying features that make the sample representative – age, gender, race, income, class, location, employment status, marital status – you could go on forever.

In either the random method or the panel method, the pollsters may make adjustments to their findings so that they become more representative. If you don't have quite enough older people in your sample, for example, you might give the answers that you did get from older people greater weight. The problem with this is that there is a danger that the pollster will start adjusting the results towards what they are expecting to happen, as we'll discuss in the next chapter on political polling.

Is the organisation making reasonable claims based on the results?

If the results of a survey claim to show that the world is massively different to how you thought, it is almost certainly because there is a mistake in the survey.

Even the biggest surveys are imprecise and have margins of error, which will be discussed in Chapter 9.

That needs to be reflected in the language used. Surveys do not 'show' or 'prove' anything, they 'suggest' things.

Looking deeply into the methodology may sound like a lot of effort to go to when you're just trying to find out if an alluring figure you heard on the news is reasonably likely to be true. So for those of you who don't have the time, there are a few quick things you can do to check.

A good start is to think about whether it's likely the people who conducted the survey would have been able to find out what they claim to have discovered. Is it something people are likely to be happy talking about? Now, check if it's easy to find out how the survey was conducted. And look at whether they seem to have gone to any trouble to make sure their sample is genuinely representative. Now, try explaining in a sentence what they have done to get these figures. If you can say it out loud without feeling ridiculous, that's a good sign.

The biggest thing to look out for is self-selecting samples. It's very tempting when you have lots of customers visiting your website, or lots of people reading your magazine or lots of followers on Twitter, to ask them what they think and publish the results. As long as you show your workings, that's fine. You might be interested in which player visitors to the Boston Red Sox website think was the best last month, but the editors presumably would not suggest that was representative of what the whole country thought. When a police force conducted a survey among its followers on Twitter, asking what they thought of spit hoods, who did they think it was representative of? It turned out that 93 per cent of the 1300 people who voted thought that the devices that were meant to prevent suspects spitting or biting were a good thing (and this was widely reported) but are followers of the police force's Twitter account likely to be a microcosm of the population as a whole? I suggest not.

To sum up the five questions, let's consider the questions you would ask about a recent finding that 56 per cent of three- and four-year-olds own their own connected device such as a tablet, PC or mobile phone. The figure just for tablets is 47 per cent. Instinctively, this feels high, especially because similar research the previous year from the regulator Ofcom found that 21 per cent of that age group owned a tablet. So let's look a bit further.

1. Where has the survey come from?
   It comes from a market research organisation that sells reports giving insights into children and young people. In this case, it's part of a big report into the use of media by children. They appear to have conducted the research themselves and there is no obvious reason for them to be biased either way.

2. What questions does the survey ask?
   The question asked was, 'Does your child have their own personal device from the following?' and then there was a list of devices. So no ambiguity there.

3. How many people were asked?
   The survey was completed by 500 parents of three- and four-year-olds. Where parents had more than one child in the correct age group the software was clear regarding which child they were supposed to be answering the questions about. The sample size is a bit small, but not small enough to explain the difference between 47 per cent and 21 per cent having tablets. Ofcom's survey was done by 600 parents.

4. Were they the right people to ask?
   Parents were found via one of the UK's largest online panels and they were chosen in order to have balance between mothers and fathers of boys and girls, and also across socio-economic groups. Ofcom, on the other hand, interviewed people face to face. It's possible that an online survey would capture more technophile parents or that some parents who could be embarrassed to admit to an interviewer that their three-year-old owned a tablet would be happy to say so online. But it's hard to imagine that either of these would have a huge effect.

5. Is the organisation making reasonable claims based on the results?
   It's tricky. Two sets of researchers have gone to considerable lengths to find the answer to a question and have reached very different figures. Neither method is perfect, but the flaws do not appear to explain the disparity. Sometimes you just get funny results from surveys, which is why you should never attach too much weight to their findings. It would probably be best not to refer to either figure on its own or to make very strong claims about either until you find a reason why they are so different.

Now you know about the five questions you need to ask to avoid being misled by surveys, you are ready to see what you are up against.

My favourite dubious surveys

I once suggested in a talk at the Royal Statistical Society that it should have a special section for people who make up numbers. I imagined that it would be like the marketing department in the company where the comic-strip character Dilbert works, where it's just one long party and every Friday they barbecue a unicorn (apparently that's a good thing). Are they having more fun than I am? I suspect their working lives must be the same as everybody else's – mostly humdrum with occasional days of excitement. The same may be said of the emails I receive from public relations companies every day. Most of them get deleted straight away, but every now and then there is a pure act of creativity that sets it apart from the usual nonsense and wins it a place in a special folder on my email system. It is some of the contents of that folder that I am now going to share with you. I will avoid mentioning the companies involved, either to avoid shaming the guilty parties, or because, despite their genius, I still don't want to promote their products.

Big number costings are a mainstay of press releases. Scrolling through my special folder I am told that people not having enough sleep is costing the economy £37 billion a year, which is coincidentally exactly the same amount as bad customer service is costing companies. Lack of neighbourliness costs the economy £14 billion a year, the skills shortage costs UK businesses £2.2 billion a year, people watching the Olympics when they're meant to be working will cost the UK economy £1.6 billion, financial crime costs the UK economy £52 billion even though hardly any of it is ever reported, having unhealthy employees is costing £57 billion, which is a touch more than the £56 billion a year they lose to poor training, and absenteeism costs the UK economy a nice big round £100 billion a year.

Looking further afield, muscle and joint pain cost European economies up to 240 billion euros a year, solar storms could cost the US economy $40 billion a day and the German national brand (whatever that is) was devalued by $191 million by the Volkswagen diesel scandal. While we're looking at the big numbers, it turns out that clumsy Brits lost £3.2 billion worth of wedding and engagement rings over the last five years.

There will be more about why costings in general are bogus in Chapter 3 and the dangers of meaningless big numbers in Chapter 6, but you can probably spot a few flaws with these ones even before you have read those chapters.

How about the research that found that 43 per cent of car accidents happen during rush hour, a figure described as 'almost half' in the press release? That doesn't seem particularly surprising, given that you would expect more accidents to happen when there were more cars on the road. But if you look in the footnotes, it turns out that they're counting rush hour as being weekdays from 6 a.m. to 10 a.m. and then 4 p.m. to 8 p.m., and suddenly it seems extraordinary that such a small proportion of accidents happen during those periods. The headline should be 'Staggeringly small number of accidents happen during rush hour'.

On the list of surveys that make me wonder how they got people to admit to things comes the survey from a recruitment company that found 37 per cent of people had admitted lying on their CVs. Be particularly careful when participants are being asked questions to which they are unlikely to give an honest answer. I was entertained by a press release that said: 'A third of people in the UK will not give truthful answers about themselves when asked questions by pollsters, according to a new survey.' In paradoxical terms, that is well up there with the words of Psalm 116: 'I said in my haste, "all men are liars".'

I love an obviously self-serving piece of research such as the survey from a company that makes technology for controlling office buildings that found 84 per cent of workers felt their productivity was seriously stunted by their inability to control the office temperature and 81 per cent would consider moving jobs for a more technologically advanced office.

A company that makes hearses reported that 91 per cent of people don't know what to do when encountering a funeral procession. But it was not a survey of the general public, it was a survey of funeral directors, 198 of them to be precise. And 91 per cent of them said people were either 'not fully aware' of what was expected of them or didn't know at all. That doesn't mean 91 per cent of people don't know what to do, it means 91 per cent of funeral directors think people don't know what to do. And maybe not even that, because it's a pretty small sample.

I must end with research that found that dogs watch around 214 hours of TV per year – equivalent to more than two years of *EastEnders* – with BBC One being the most-watched channel. Sometimes you wonder not just about the methodology, but about the point.

Avoiding being misled by surveys is all about training yourself so that alarm bells go off when you hear about them. I hope yours have been jangling throughout the last few pages.

Take as a starting point that surveys are guilty until proven innocent, then apply the five questions and you should be able to live your life untroubled by survey-induced nonsense.

CHAPTER 2

Opinion Polls

Should you believe them?

Let's take what you now know about surveys and apply it to political opinion polls. Around election times they generally ask which party people are planning to vote for. Which party is leading in the polls becomes a key part of the narrative around election campaigns, so there are even fresh polls straight after a televised debate to see which party has benefited the most from it. Polls can be interesting, but it's important not to accept them as gospel truth and it's especially important not to take too much notice of a single poll.

This is why: two weeks before the referendum on Scottish independence in 2014, the *Sunday Times* published a YouGov poll, which put the 'Yes' campaign ahead by 51 per cent to 49 per cent, excluding the 'don't knows'. Seemingly as a direct result of this, Chancellor George Osborne announced that a timetable would be set out for giving more powers to the Scottish Parliament if there was a 'No' vote, including more powers to raise taxes. The following day the news was knocked off the top spot by the announcement that the Duchess of Cambridge was expecting her second child, but former Prime Minister Gordon Brown set out the timetable for devolving more powers. The government in Westminster denied it was panicking but announced that instead of attending Prime Minister's Questions, Prime Minister David Cameron and Labour leader Ed Miliband would be heading to Scotland to campaign for a 'No' vote. Deputy Prime Minister Nick Clegg would also be travelling north to campaign, although none of them would be doing so together.

All this created challenges for the BBC News coverage because BBC editorial guidelines say that news programmes should not be led by the results of a single opinion poll. On the other hand, the government was clearly reacting to the results of the single poll, so news stories had to start with what the government was doing and mention the poll later on. It's a fine line to tread.

The day after the poll was published, I wrote a piece on the BBC News website called 'The perils of unprecedented polls', which discussed why it is so much more difficult to conduct meaningful opinion polls in one-off elections than it is for regular ones such as general elections.

In the end, Scotland remained part of the UK by 55 per cent to 45 per cent. There may have been other things going on, but it certainly looked as if the results of a single poll had led Cameron, Miliband and Clegg to drop everything and head to Scotland for a final push and to offer extra powers to the Scottish Parliament if the Scottish people voted to stay in the UK.

It's possible that on that Sunday the 'Yes' campaign was in the lead and drastic measures were needed. Alternatively, it may be that there was an overreaction to a 'Yes' lead within the margin of error when other polls still had 'No' ahead. We'll never know for sure, but worrying too much about a single poll is almost always a mistake.

This chapter will look at opinion polls around election times, and why you need to be careful with the results you see. In particular, I'll look at the key questions you need to ask that will help you work out what you can read into the poll:

- What can we learn from previous elections?
- How are exit polls different to regular opinion polls?
- What is the margin of error for an opinion poll?

What can we learn from previous elections?

Much of what you need to know about opinion polls is the same as the issues around surveys. In this case, you are trying to predict what will happen when everybody in the country is asked a question (but doesn't necessarily have to answer), i.e. you are trying to forecast the result of an election. Because you can't afford to ask everyone the question in advance, you ask a smaller number of people and hope they are representative of the people who are going to vote in the election.

You will generally see opinion polls asking at least 1000 people, and there is once again a key division between companies that carry out random sampling and those that have online panels. The difference is that the random samplers try to find ways to choose people such as randomly picking numbers out of the phone book. Pollsters using panels have thousands of names in a database of people who are prepared to complete polls for them and about whom they know a great deal. That means when they are asked to conduct a poll they can find a group of people they think are representative of the population. In both types of polling there can be adjustments made later on by the polling companies, and I will return to the challenges that this creates.

These methods have been developed largely through trial and error, based on the experiences of some of the classic mistakes in the history of polling. The first key case study is the 1936 US presidential election, when the incumbent Franklin D. Roosevelt was up against the Republican Governor of Kansas, Alfred Landon. One of the country's most widely respected magazines, the *Literary Digest,* which had correctly predicted the outcome of several previous elections, decided to conduct an enormous poll and got responses from about 2.4 million people. Just pause a moment to consider how large a poll that is – one of the largest ever conducted. But even that understates the process, because they sent dummy ballot papers to 10 million addresses. The amount of work involved is mind-boggling.

The weekly magazine created its mailing list based on lists such as all the telephone directories in the USA, country-club memberships and car registrations. It sent a ballot paper to each person on the list and asked them to complete and return it. A quarter of them did so, meaning that staff had to open and record the contents of 2.4 million letters. Based on those responses, the *Literary Digest* confidently predicted a victory to Alfred Landon by 57 per cent to 43 per cent. If you're wondering why you have heard of Roosevelt and not Landon, it is because the forecast was a staggering failure from one of the most expensive opinion polls ever conducted. In the event, Roosevelt won by 62 per cent to 38 per cent in one of the biggest landslides in a presidential election. It's a classic example of the fact that it doesn't matter how many people you ask if they're the wrong people.

Remember that this was the end of the Great Depression; there were still about nine million people unemployed in the United States, and owning a telephone or a car was a significant luxury. Making up the mailing list from phone books, country-club memberships and car registrations skewed the sample towards more affluent voters. It ignored those who were most likely to have been helped by Roosevelt's New Deal, which had brought the country out of the depression.

Also, only a quarter of the people who received the ballot in the post bothered to return it, and the people who could be bothered to return the ballot in the mail turned out not to be representative of the electorate as a whole.

The outcome of the 1936 election had been correctly predicted by George Gallup, who had used a quota system and a very considerably smaller sample, but he came unstuck in the 1948 presidential election between New York Governor Thomas Dewey and incumbent President Harry S. Truman. The idea of the quota system was to select characteristics of the population such as race, gender and age, and then make sure that the sample included the right proportions of such people. To get the sample of 3250 people polled, a professional interviewer would be told, for example, to find ten black women under the age of 40 living in a city. Beyond that, the interviewers were allowed to choose the people themselves. That element of human selection biased the sample again, with Gallup predicting a victory for the Republican Dewey by 50 per cent to 44 per cent (with the rest going to third party candidates) when in the event Truman won by 50 per cent to 45 per cent.

To be fair to Gallup, the other major pollsters of the time also predicted a win for Dewey, and Truman was widely considered to be the underdog in the election. The *Chicago Tribune* was so convinced that Dewey was going to win that it ran a headline in its early edition saying 'Dewey Defeats Truman'. The photo of Truman holding up the newspaper declaring victory for his opponent is one of the iconic moments in US political history. All the major pollsters were using the quota system. Presumably, at the time, Republicans were easier to find and interview than Democrats. It was also suggested that the poll was conducted too far in advance of the election. In the poll two weeks before election day, 15 per cent of the sample were undecided and it was assumed that their votes would split in the same way as those who had already decided. But Truman was good at campaigning in the last few days, so there may have been late decisions that did not show up in the poll. I'm always a bit suspicious of this though – the easiest defence if your opinion poll turns out to be wrong is to claim that voters changed their minds on election day.

The proportion of people undecided is an important thing to check when you're reading the results of a poll – it's the Whiskas effect. One of the best-known slogans in British advertising is that eight out of ten cats prefer Whiskas. But while the original advert for the cat food said that eight out of ten owners said their cats preferred it, I've seen references to the company being forced to change the line to reflect the fact that it was only eight out of ten cats whose owners expressed a preference. I have asked the Advertising Standards Authority to find that ruling for me and they couldn't, but the owners who don't express a preference are important. For all we know, 99 per cent of owners could have said their cats couldn't tell the difference between any types of canned meat, and the agency would have had to ask 1000 cat owners just to find 10 who could distinguish between brands.

Consider a situation in which a polling company interviews 1000 people and 300 of them say they will vote 'yes', 200 of them say they will vote 'no' and the rest say they haven't decided. If the reporter ignores the 'don't knows', the headline could be 'Yes leads in polls by 60 per cent to 40 per cent', but that would not really tell the current story of the election. Reporting that 'yes' was leading by 30 per cent to 20 per cent would much more accurately get across that there was still much to play for in the campaign.

This is an important decision in all surveys. It is often tempting to assume that people who say they don't know or have not yet decided would end up being like the rest of the population once they had made up their minds, but there is rarely evidence for this. Ignoring the undecided respondents is a misleading thing to do.

In the UK, one of the most famous times that the pollsters got it wrong was in 1992, when opinion polls consistently showed Labour leader Neil Kinnock slightly ahead of the Conservative incumbent Prime Minister John Major, with a hung parliament (in which no party has an overall majority) widely expected. The Conservatives had been in power for 13 years, there had just been a long recession and interest rates were above 10 per cent. Yet in the event, Major's Conservatives got more votes than any party before or since, won the popular vote by eight percentage points and retained a small majority in Parliament.

We never find out exactly why polls are wrong, but in this case the pollsters blamed three things. They suggested there had been a late swing to the Conservatives (which I've already said I'm suspicious about – the only evidence for that would be polling data, which feels like marking your own homework). They said that people had been reluctant to admit they were going to vote Conservative because it wasn't fashionable, a phenomenon known as 'Shy Tories'. And former YouGov president Peter Kellner suggested that there were sampling errors because the results of the 1991 census, which were published shortly after the 1992 election, showed there had been a sharper contraction in the working class and growth of the middle class during the 1980s than had been thought, which meant the sampling design used by the polling companies was skewed away from the Conservatives.

The outcome of the 2015 general election, when David Cameron's Conservatives unexpectedly took an overall majority, had also not been predicted by the polls. The polls were effectively predicting a dead heat, when the actual result was 36.9 per cent to the Conservatives and 30.4 per cent to Labour. The post-mortem into the polling at the election concluded that once again the problem had been with the sampling; that the polling companies had taken samples that unrepresentatively favoured Labour over the Conservatives and that adjustments made to the raw data had not solved this problem. The report also pointed out that while there had been polling at UK general elections in the past that was nearly as inaccurate as 2015, it hadn't received as much attention because it still managed to predict correctly which party would win.

The other growing problem for opinion polls, which I mentioned in the chapter about surveys, is that you have to phone an awful lot of people to get any of them to respond to your opinion poll, with one polling organisation saying that to get 2000 responses they were having to call about 30,000 random numbers. If a one-in-four response rate caused problems for the *Literary Digest* in 1936, a one-in-15 response rate must be a cause for concern. The people who are responding are unrepresentatively interested in talking to pollsters, which may well be introducing bias into the survey in other ways.

If the catchphrase to explain the 1992 errors was 'Shy Tories', in 2015 it was 'Lazy Labour', with the suggestion that people telling pollsters that they were going to vote Labour were less likely to turn out and vote than people who said they were going to vote Conservative. The pollsters also overestimated the likelihood of younger voters turning out. These are the sorts of things for which polling organisations try to adjust once they receive the raw data. It may be that you have heard from too few people of a particular age group or too few in a particular part of the country, for example, so you adjust your findings to reflect that, putting more weight on the responses you have received from under-represented groups. Then adjustments can be made for whether you think people are actually going to turn out to vote and even whether you think they are telling you the truth. The trouble is that such adjustments are based partly on experience from previous elections and partly on what the polling organisation thinks the outcome is likely to be. That may encourage group thinking among the pollsters, who could be tempted to adjust their results to make them tally with what their competitors are finding. That was not the case in 2017, when there was a much wider range of outcomes in the polls even if they did, on the whole, overstate the support for the Conservatives and underestimate Labour's support.

The experience from previous elections is important, which is why there are such severe challenges in one-off elections such as referendums. While you can have a pretty good guess how big the turnout will be in a general election, it is harder to predict, for example, how many people will vote on whether the voting system should be changed. It is also harder to know whether there is anything going on such as the Shy Tories or Lazy Labour phenomena in a referendum – is it going to be unfashionable to vote 'no' to Scottish Independence or 'yes' to Brexit? Also, as there are generally only two possible outcomes in a referendum, if you predict the wrong one, even if the result is really close, you are going to look bad.

I say there are only two possible outcomes, but I was asked about a possible third one when answering questions with the Reality Check team from listeners to BBC local radio stations during the EU referendum campaign. We spent a day in a radio studio in Westminster, answering questions for 20 minutes per station. It was fascinating, and the best question I was asked was what happens if the result is a dead heat. In an electorate of 46.5 million, of whom 33.6 million voted, it is staggeringly unlikely that there would be a draw, but staggeringly unlikely things do happen. I didn't know the answer so I asked to move on to the next question and said I would get back to them with an answer. But the answer is that there isn't one – there is nothing in place to answer the question of what happens if there's a draw. As the EU referendum was, strictly speaking, not binding, the government would just have had to decide what to do, but it would have been very embarrassing.

We know what happens if there is a draw in a local election – there was one for a seat on Northumberland County Council in 2017. In such circumstances it is down to the returning officer to decide how to break the tie. On this occasion, after two recounts, they decided to draw straws, but they could have chosen any random method such as tossing a coin or drawing names out of a hat. There are no such procedures in place for referendums.

In a general election, the prime minister is the person who can command a majority in the House of Commons, so a tie would not be a problem in the same way. Understanding the way an election works, whether it is just who gets the highest proportion of the popular vote or a first-past-the-post constituency system, is important.

How are exit polls different to regular opinion polls?

You may have noticed that there was not an exit poll at either the Scottish independence referendum or the EU referendum. An exit poll is different to a normal opinion poll because it is conducted by approaching people outside polling stations after they have voted. This means that the pollsters do not have to worry about whether or not the respondents are actually going to vote. The idea is to forecast the number of seats won by each party, as opposed to the national share of the vote. The exit polls you may have seen at UK general elections recently have been commissioned jointly by the BBC, ITV News and Sky News. The polling stations where the general election exit poll is conducted are carefully chosen based on past experience as the ones that give the best indication of which way the country as a whole will be voting. They speak to thousands of people (16,000 in 2010, for example) across a tiny proportion of the 39,000 polling stations across Britain – about 140 of them. It's a much bigger sample than you have in most opinion polls, but remember the lesson from 1936 that it doesn't matter how big the sample is if you're not getting the right people. The exit poll tends to be conducted in marginal seats, which change hands the most often, because that is where governments are made or broken.

The recent record of exit polls is pretty good, although they were wrong in predicting a hung parliament in 1992, just as the regular opinion polls had been. They had been less accurate in 1987, but still managed to predict the correct winner. The exit poll has done well in the last four general elections, getting the Labour majority right in 2005 and predicting a hung parliament with smaller support than expected for the Liberal Democrats in 2010. In 2015, it predicted much stronger support for the Conservatives than had been expected, although even it did not foresee the overall majority, but it did predict the huge rise in seats for the Scottish National Party and the collapse of Liberal Democrat support, which had former Lib Dem leader Paddy Ashdown saying he would publicly eat his hat if the exit poll was correct. He was later given a hat-shaped cake on *Question Time*. And in 2017 the exit poll predicted the smaller-than-expected support for the Conservatives.

The point about exit polls is that they try to establish the swing between parties. This is so that they can work out whether a particular party will do better or worse than it did last time based on what they are told by a proportion of voters at polling stations that have proved to be representative in the past. If, as in the case of referendums, you do not have the historical results to build an exit poll on, it is much harder to do and the risk of calling a two-horse race the wrong way is not worth taking.

It's also important to bear in mind that exit polls only speak to people who vote at polling stations so it will not take account of people voting by post.

What is the margin of error for an opinion poll?

Because polls are based on a sample, they have a margin of error, just as surveys do. In a properly random opinion poll based on 1000 responses, you can be 95 per cent confident that the margin of error will be plus or minus three percentage points. So if the poll finds that voters are evenly split on a 'yes' or 'no' question, it's likely that the actual response is somewhere between 53 per cent and 47 per cent in favour of 'yes'. Clearly, therefore, if you are looking at a close-run contest, you shouldn't read too much into a lead of one or two percentage points, as was the case with the Scottish referendum poll mentioned earlier.

That margin of error comes down to plus or minus two percentage points if you poll 2000 people. But this assumes that the way you are sampling is accurate and that no sampling biases have crept in. If it turns out that only a quarter of people contacted actually responded to the poll, or supporters of one side in the election were much more enthusiastic to talk about it, or key groups of supporters of one side were particularly difficult to get hold of, then the margin of error could be considerably greater. Also, the process of weighting the raw data to increase the impact of under-represented groups increases the margin of error.

Strictly speaking, these margins of error only refer to polls using random samples and not panels. For a sample to be random, every member of the population being surveyed must have an equal chance of being part of the sample. With online panels we know this is not the case because not everybody has internet access and the panels rely on people opting into the process rather than being chosen randomly. Companies that conduct such polls are experimenting with alternative ways of getting across the level of uncertainty.

Also, do make sure that the poll you're reading has been conducted by a reasonably reputable organisation, which has a code of conduct that prevents the worst abuses of polling; like surveys, this includes wording questions in a way that will mislead respondents or encourage them to answer in a particular way. The polls commissioned during an election period by national newspapers will often be conducted by British Polling Council members, who should be transparent about the way they conduct their polls. Polling companies seem to behave themselves better during election campaigns when their reputations are on the line.

The BBC never conducts polls asking people how they plan to vote, but it is one of the broadcasters that commissions the general election exit poll, which asks people how they just voted.

So the message is, if you have seen several reputable polls that suggest one side is leading by about 70 per cent to 30 per cent, you can be pretty confident that side is going to win. The exit poll for the Irish vote on legalising abortion was pretty conclusive although, as discussed, conducting an exit poll was a risky thing to do. If you have seen several polls calling it as 51 per cent to 49 per cent in either direction then you would be best off concluding that the outcome is too close to call and there is not a lot else that opinion polls can tell you about this particular contest.

This was the case with the EU referendum – in the days leading up to the vote there was no consistent picture, with some polls giving Remain a slight lead, some giving Leave a slight lead and some having the two neck and neck. It is frustrating that polls aren't more helpful in a close contest, but they are not and it is important to recognise that they are not. That is why BBC Editorial Guidelines (which are publicly available to read online) indicate that staff should be suspicious of them.

The lesson from historical elections is that taking accurate samples to test opinion during an election campaign is difficult, with pitfalls including asking loads of people but excluding big sections of society; letting pollsters choose who they want to interview; and ignoring ways in which supporters of each side may differ, such as whether they are more or less likely to tell you for whom they are going to vote or whether they will turn out on polling day.

Knowing how the poll was conducted helps you test whether it was robust and whether the questions asked were sensible and clear. Also, make sure you know what the organisers have done about undecided voters – just ignoring them and assuming they will end up being the same as the rest of the population is misleading.

Finally, try not to let your own decision on whether or how to vote be influenced by polls, especially ones where the gap between the parties is close to the margin of error – you will regret it if they turn out not to be accurate.

CHAPTER 3

Cost

Bear in mind that costings are bogus

Picture the scene: it's a bit snowy in London. Not snowy in a way that people from places with proper weather such as Canada or Norway or Scotland would be concerned about, but there are a few inches of snow on the ground.

Parts of the transport system have ground to a halt, but you have nonetheless managed to fight your way into work. Whatever you may think of journalists, imagine you have walked miles through the snow to get to the office of the newspaper where you work and you have made it just in time for your morning editorial meeting.

Your editor, a wizened old hack, stares out of the window and says: 'It's pretty snowy out there. I bet that's costing the economy a pretty penny.' And he turns to you and tells you to go and find out how much it has cost.

This has been the experience of many a reporter. It is a knee-jerk reaction for editors to ask how much things cost. Let's look at an example from the *Daily Telegraph* on 2 February 2009, not because it's any worse than many other versions of this story, but because it shows its workings well.

It has the headline: 'Snow Britain: disruption could cost UK economy £3bn'. The figures come from the Federation of Small Businesses (FSB), which has apparently warned that the snow would make the UK economy lose out by £1.2 billion on Monday and Tuesday, with smaller losses in the rest of the week, making the total up to £3 billion.

How has that figure been worked out? The article explains that the FSB: 'made the calculation based on the assumption that 20 per cent of the workforce or 6.4 million people were off work on Monday because of the weather conditions, and that an average bank holiday costs the UK economy £6bn'.

The economists think that a bank holiday costs £6 billion, because they made that number up previously – we'll come back to it – and they know that nobody works on bank holidays at all, which may come as a surprise to any journalist or indeed nurse, police officer, supermarket worker, transport worker . . . It's not clear where the figure of 20 per cent of people not making it to work comes from, but the economists have decided that one-fifth of people not making it to work will mean the snow will cost 20 per cent of the amount a bank holiday costs.

Do we believe this? The cost of snow is an excellent example of the dubious costings that editors seem to love. I've been fighting against them for years. In this chapter we will be looking at why any headline showing the cost of something is likely to be bogus. Remember, I'm talking about the cost of things, not the price. There is no question that if I go and buy a chocolate bar at the local shop I will know what the price is – it's there on the shelf. But how much it has cost to get that chocolate bar on the shelf is another question and it is not an exact science.

This chapter will go through three things you need to understand whenever you see figures for cost:

- What is meant by cost to the economy?
- Are we talking about the total cost or the extra cost?
- Questionable costings in business

What is meant by cost to the economy?

Back to the cost of snow. When we left our calculation, we were considering whether £6 billion is a reasonable costing for a bank holiday. If you added up everything produced in the UK economy to generate a figure for gross domestic product (GDP), you would get about £2 trillion a year. There are about 252 working days a year, so that's getting on for £8 billion a day, although it was closer to £6 billion a day when the *Telegraph*'s article was published in 2009.

Does it make sense to see a bank holiday as an almost total loss of output to the economy? One of the fundamental problems of the UK economy at the moment is low productivity – that's the amount produced by workers per hour – so maybe giving people a day off every now and then would help them work more effectively the rest of the time. Also, bank holidays are a jolly good thing for some parts of the economy. If you're selling ice creams on Margate beach you would laugh at the idea that having a bank holiday would lose your business money.

So, we've started with a dubious number, divided it by five, and that's how much snow has cost the economy. But there's an even more serious problem with this calculation, which is the assumption that snow is bad for the economy at all. Just as bank holidays are probably going to be good for people selling ice cream, so snow also has its economic upsides (though probably not for ice-cream sellers).

When we talk about things being good or bad for an economy we are talking about whether they will increase or decrease the level of GDP, which is the measure of the total amount produced in the economy. It's clear that there are some parts of the economy that will lose out as a result of the snow. If you are running a car plant in Sunderland on three shifts a day and you lose a shift because your workers cannot get in or parts cannot be delivered, then that is going to be a dead loss for you.

But not much of the UK economy is like that – 79 per cent of the UK's GDP comes from the service sector. If you are a hairdresser and you cannot open on Monday or Tuesday, the people whose appointments were cancelled will still need haircuts. You may need to work a bit of overtime in the following week or so, but you are not going to lose the whole two days of takings. More and more people can work from home, even if their children's schools are closed due to snow. It's not that people failing to get to work isn't a problem for the economy – if you're selling sandwiches to office staff in a city centre you will be likely to make less money that day, and your customers will be unlikely to eat extra lunch in the coming days to make up for it. But it's not as much of a problem for the economy as it was when the UK was more dependent on manufacturing.

Now look at the benefits to GDP from snow. If councils pay people extra to spread grit and clear roads, that's a boost to GDP. In fact, some of the rock salt that is used to grit roads is mined in places such as Cheshire and County Antrim, and having lots of that bought by councils is particularly good for the economy.

Also, if people have to stay at home there are more opportunities for online shopping, which will boost the economy. Sales of winter clothing would be expected to increase, and the extra spending on heating during the cold spell would also be good news for GDP.

Imagine people crash their cars in the icy conditions. They might take the car round to the local garage and pay to have it fixed. If an insurance company has to pay out that would be a transfer of money that might otherwise go to shareholders who would be marginally less likely to spend it, so that's good for the economy too. And if the car is written off and the owner uses the insurance to buy a new one, well maybe that will help the factory in Sunderland to cope with losing a shift.

I have occasionally talked on my statistics courses about the benefits to the economy of increased demand for replacement hips when people slip on ice. The manufacture of replacement hips is a great British hi-tech industry. I have been accused of being callous, which is missing the point. What is good for the economy is not necessarily good for people.

You could suggest that all this is more of an argument for why GDP is a bad measure of what is going on in the economy than anything about snow. If you crash your car and have to pay someone to return it to its previous condition, should that really be seen as a benefit to the economy? The point is that some aspects of snow boost GDP and some are bad for it, and it's very hard to tell on the day the snow falls which will be the case.

If snow had a significant effect on the economy then you would expect the Office for National Statistics (ONS) to mention it when it publishes its quarterly GDP figures. In the first quarter of 2009 it didn't mention it at all, so we can assume that the snow did not cost the economy £3 billion.

The following year, there was heavy snow in the week before Christmas, and that was a big deal – a perfect storm if you like. People could not get to the shops to buy gifts, and office parties in bars and restaurants were cancelled. That delay in purchases was significant because if you buy things in the week before Christmas they are more likely to be at full price, whereas if you buy them after Christmas they are more likely to be discounted (as well as too late).

Also, because the snow was right at the end of the year, even money lost to the economy that was later made up (i.e. people going to the hairdresser the following week) would have gone into the GDP figures for the first quarter of 2011. They could turn up even later – the part of the ONS that calculates the GDP figures delayed its own Christmas party until the following April, so that would have turned up in the figures for the second quarter. It's not often that Christmas parties boost second-quarter growth.

For the last quarter of 2010, the ONS did indeed acknowledge the damage done by the snow. It said that the weather had knocked about half a percentage point off that quarter's GDP, which is a considerable hit. Half a percentage point off a quarter's growth is just over £2 billion, so even the weather event that was bad enough to be mentioned by the ONS did not reach the £3 billion mark.

In 2018, the UK once again had enough snow in London to dominate the national news agenda for a few days, when the 'Beast from the East' swept in from Siberia. The front page of the *Observer* on 4 March carried the headline: 'Freezing weather costs UK economy £1bn a day'. That's a nice, round number, exactly the same as the amount that the *Daily Telegraph* said it was costing in December 2010 in its headline: 'UK snow: bad weather costing economy £1bn a day'.

That's actually not a great coincidence as both figures came from the same source, the Centre for Economics and Business Research (CEBR) in 2010 and its founder Doug McWilliams in 2018. He tweeted that total output each day would be cut by 20 per cent, even after the effects of online shopping, working from home and a 20 per cent increase in energy output were taken into account. He described it as 'a very rough estimate' and told me he expected a lot of it to be made up by the end of the quarter this time round. But if it's going to be made up then it's not being lost to the economy at all – delayed spending is still spending.

Nonetheless, for only the second time, the ONS did indeed mention snow in its GDP report. Its report said: 'While some impacts on GDP from the snow in the first quarter of 2018 have been recorded for construction and retail sales, the effects were generally small, with very little impact observed in other areas of the economy.' So still not £1 billion a day then.

The problem here is not so much the way that the FSB or indeed the CEBR went about calculating the cost of snow, it's that the question was asked at all. When snow falls there seems to be a strong urge in the minds of editors to think about costs to the economy, despite the lack of evidence that there is one. It is extraordinary that a single tweet from a single economist would be enough to create the front-page headline on a national newspaper, especially when it was the same headline run by a rival newspaper eight years before.

This is not just about snow, it goes for any events and their cost to the economy: fires, earthquakes, train strikes – we do not know on the day how much they will cost the economy. I was particularly shocked at the interviews being done after the Asian tsunami on Boxing Day 2004, when questions were being asked about how much it would cost the economy when we were not even close to an accurate figure for the hundreds of thousands of people who had been killed. The cost to the economy was not the story. Also, as aid pours into relatively poor parts of the world to fund reconstruction work, the economy gets a boost, which may even make up for the immediate loss of economic output at the time, but that does not make it any less of a disaster.

In the case of events that are important to the insurance industry such as flooding in the developed world, the industry may come up with an early estimate of how much it expects to pay out, which is a more useful number than the cost to the economy, but it's still a very rough estimate.

Figures that you hear on the day of an event for how much it is costing the economy are not reasonably likely to be true. That's the case wherever you are in the world, even places where snow is expected and planned for.

It's not as if there aren't numbers available that would make more sense. The owner of the café on the corner could tell the nation how much they reckon they have lost as a result of people not getting into work. Maybe they would say how many fewer coffees they have sold than they would normally have done by that time of day. That number of coffees has the advantage of probably being an accurate figure, allowing the audience to get a handle on the idea that some people are losing money and not being a meaninglessly big number such as £3 billion, which the brain can't really process anyway. And also it's not just been plucked out of the air.

Are we talking about the total cost or the extra cost?

Bogus costing is not just a problem in news headlines. To illustrate why, consider what happened to me in Egypt in 1990.

When I was 16 I went on a walking tour of the Sinai Desert with the Israeli nature protection society. I had a great time until about four days in when, at the bottom of a valley, I slipped down a gap between two boulders and broke my right ankle very badly. While forms of mobile communication did exist, they were banned in Sinai by the Egyptian authorities, presumably in case anyone wanted to spy on their desert. We sent part of the group off for the four-hour walk to the nearest telephone and another part walked off to the next campsite to bring back supplies. We had a medic with us, but the strongest painkiller he had was paracetamol, which wasn't really strong enough for the situation.

When the group reached the telephone they called the United Nations at Taba, on the Israeli border. They said they would love to come and collect me, but I'd clearly need winching out of the valley and they didn't have any helicopters capable of winching. The only people in Sinai who did have such helicopters were the Multinational Force and Observers (MFO), which was a set of peacekeeping troops in southern Sinai set up as part of the Camp David Agreement. My friends suggested the UN might like to call the MFO, but the problem was the UN did not recognise the MFO, so a French major in Taba had to call a Russian general in Cairo to get permission to call the MFO down the road.

They eventually got permission and the MFO said they would be delighted to come and rescue me, but it was getting dark and it would be a dangerous mission at night, so they would turn up in the morning.

The following day, 19 hours after I'd broken my leg, two helicopters turned up at my valley. They had needed to call out a second one because, by the time they had found me, the first helicopter didn't have enough fuel left for winching me out. Two US Army medics were dropped off at the top of the valley with a metal stretcher and sent down to strap me into it. They introduced themselves as 'Bob and Dean from Pennsylvania and Alabama, that fine southern state'. I was obviously pleased to see them, and as a Royal Air Force cadet I was beyond excited at the idea of being winched into a US Army 'Huey' helicopter. Bob told me to keep my eyes shut as I was winched so I didn't get sand in them, but that was never going to happen. It was an odd sensation, which reminded me of the bit in the first Superman film when our hero catches Lois Lane in mid-air and says, 'Easy miss, I've got you,' and she says 'You have got me – who's got you?' When I got into the helicopter, all the soldiers thanked me for giving them something interesting to do that day. Apparently I'd got them out of a boring training exercise. It takes a very special person to save your life and then thank you for the opportunity to do so.

If you have ever watched a Vietnam War film you'll know what a Huey – the nickname for a Bell UH-1 Iroquois – is like. In particular, the sides tend to be open. I was perfectly safe because my stretcher was attached to the floor, but nobody had mentioned that to me, so the banked turns were a bit scary. They took me to an excellent hospital in Eilat in the south of Israel, with the only further incident being that we were aggressively challenged by a US Navy battleship that described us as an unidentified aircraft. This was summer 1990 and the Gulf War was just kicking off, so there was a certain amount of nervousness in the area. The pilot who said we had been challenged by the US Navy also said that everyone should duck.

So why am I telling you my teenage adventure story in a book about statistics? Well, when I got back to school, somebody asked how much the US Army had charged for rescuing me. When I told him the army hadn't charged at all, he said that was very generous. Don't get me wrong, I owe my life to the wonderful servicemen who rescued me from that desert (not to mention my tour group who so willingly helped in so many ways), but how generous was the US military being?

Let's try making a shopping list of what it cost to rescue a foolhardy British tourist from the middle of the Sinai Desert. You start with the cost of two helicopters for a day, which is seriously expensive. I know that because my father had been rescued by helicopter after a skiing accident three years previously (my poor mother . . .) and our insurance did have to pay for that one. That rescue and treatment almost broke through the limit for our travel insurance, so we know helicopters are properly pricey – my family is very much in profit over travel insurance providers. On top of the helicopters themselves, we needed at least three tanks of fuel, because the Huey I was in had to stop to refuel near Eilat so it could return to base. Then there was personnel: a pilot and co-pilot for each helicopter, plus Bob and Dean, and there were at least two other soldiers looking after me in the back of the Huey, so that's at least eight highly trained people, working for most of the day, not to mention the support staff back at base who kept them in the air. And they used a whole lot of splints, bandages and drips when they were patching me up. If you add all of that up you have to be getting into six figures for the cost of saving my life. Well worth it, I'd say, but would the US taxpayer agree?

Now, let's think about this another way. All of the staff involved were going to be working that day anyway. The funding of the MFO is established by international treaties and I'm not sure there's any allowance for some of it to be refunded by the insurance companies employed by British teenagers. If they hadn't come to rescue me that day, the helicopters and fuel were due to be used up in an exercise. Instead, they were used on a mission, which from talking to the soldiers seemed to have improved morale – apparently being posted to Sharm El Sheikh was a bit boring. So the only extra cost to the US taxpayer was some medical supplies.

Depending on how you think about it, the cost of rescuing me was either hundreds of thousands of pounds or almost nothing with a boost to troop morale thrown in. These two ways of thinking about the cost of something are known as the total cost and the marginal cost. The marginal cost is the extra cost of doing something so, in this case, it's the cost of a few medical supplies. The total cost is the hundreds of thousands of pounds it would have cost to rescue me out of the desert if all the equipment and personnel and fuel hadn't been going to be used that day anyway.

This is an important distinction and you'll spot it in the news all the time. A regular favourite is how much the security is going to cost for a demonstration. You may well need 100 police officers to attend, but what would they have been doing if they hadn't been doing that? Are they all coming in on their days off and being paid overtime, or is covering marches just part of their job? Should you take the total cost of the barriers and other equipment used and divide it by the number of times you think the equipment will be required, or do you just decide that's something the police have to own anyway? If you wanted to make a particular police operation sound expensive it would be easy to pile up the costs, but you could also make it look negligible if you wanted, and that is why all figures you see for costs are at the very least questionable.

We saw estimates of anything from £2m to £30m for how much the security cost for the Royal Wedding in 2018 between Prince Harry and Meghan Markle, but bear in mind when you see those that police forces can apply for extra funding from the Home Office for dealing with events that are outside their usual remit. Thames Valley Police, which provided security for the event in Windsor, is not awash with cash, so it could be forgiven for billing for its total costs rather than its marginal costs. But that does not mean it is reasonable to say that the event cost the police that amount.

Every now and then we get a figure from the National Health Service (NHS) for how much each missed appointment costs it. In January 2018 we were told by the *Guardian* that missed appointments cost the NHS in England a nice round £1 billion a year. That was based on eight million outpatient appointments being missed at a cost of £120 each. The money could instead have been spent on 250,000 hip-replacement operations, we were told. To be clear, it's important to turn up for appointments in the NHS, but that doesn't make it OK to use misleading stats.

The £120 per appointment figure comes from taking the total cost of having outpatient services in the NHS and dividing it by the number of appointments. But for a missed appointment to cost the NHS £120 you have to assume that all the doctors, nurses and support staff will be sitting around doing nothing during the scheduled time for that appointment (although they will still be consuming the sort of single-use equipment that will come under the average cost). In reality, the NHS will almost certainly bear in mind when scheduling appointments that a certain number of people will not turn up. And even if they didn't, I can't remember the last time I saw a doctor or nurse who didn't look like they were doing anything useful and couldn't use a few minutes to catch up with themselves. So missed appointments probably don't cost the NHS very much at all – in fact, it would probably struggle to cope if everybody did turn up. It certainly wouldn't be able to pay for an extra quarter of a million replacement hips.

Costings are dangerous things because it is easy to make something look cheap or expensive depending on how you work them out. Costings are most often used to make things look expensive, so whenever you see an impressively big price tag being attached to something, just ask whether any of the costs would have had to be paid anyway.

Questionable costings in business

The ambiguities around costings also affect the world of business. Costs are clearly a crucial figure on any balance sheet. Imagine I have a company producing cuddly seal toys. I have rented some space on a factory floor, bought some sewing machines, taken on some machinists and bought in supplies of stuffing, cotton, synthetic fur and plastic features. Now I need to work out how much it costs to produce each cuddly seal. Much of this is obvious – I know how much I have paid for the amount of material I need for each seal. I have a good idea how long it takes each of my employees to make a seal and I know how much I am paying them per hour. But then there are trickier questions. I don't really know how many seals can be made with each sewing machine before it wears out so I need to guess. I could decide that I would expect a sewing machine to last a year, or five years, or ten years. Whichever one I choose will make a big difference to how much of the cost of the machines I add to the cost of each seal, which means that it makes a big difference to how much profit I make on each seal.

These calculations go into my company's management accounts, which is what I use to run my company. The other set of figures are my financial accounts, which is what I publish, using various complicated accounting regulations and arbitrary assumptions to try to make my company's results comparable with other businesses. These are the figures you will read about when companies publish their annual results. Companies have auditors to prevent them doing misleading things in their financial accounts, but even within what is allowed there is considerable wriggle-room. For example, in the financial crisis, financial institutions were giving very different valuations to the same assets, even though they were often being audited by the same big accountancy firms. There have also been famous cases of now-bankrupt companies having fiddled their figures by declaring items that should clearly have been operating expenses as some sort of multi-year investment.

My cuddly-seal-making company is very simple, so it's pretty clear when you take a step back whether the way I'm running it is reasonable. But most companies are more complicated than that, and the management software and accounting systems that they now use make it more difficult to spot things that in retrospect are clearly misleading.

There is no question that there are ambiguities in costs and valuations for companies. Consider the example of a company that owns a railway carriage. What value do you give the carriage in your accounts? It could be the amount that it would cost you to replace it with a new one, or at least a second-hand one, which would make it a valuable asset. Or you could take into account how much it would cost you to scrap it at the end of its life, which could make it a liability.

The costs really matter when a business is trying to take decisions. Imagine that the zoo is getting some polar bears, and they put out a tender for cuddly polar bear toys. My company producing the cuddly seal toys decides it might diversify and bids for the contract. How much does it cost me to make a cuddly polar bear? More of the synthetic fur, stuffing and plastic features will be needed, so I definitely have to include them in the cost. And I'll need to hire a couple of extra staff, so that is part of the cost. But there is time during the day when not all my sewing machines are being used, so they can be used to make the polar bears – do I include a proportion of the costs of the sewing machines in my final figure? And what about me? I'm having to devote some of my time to this, so perhaps I should allocate a proportion of my wages to the cost of the polar bears. I won't need to get any more factory space or spend any more on heating or lighting, but should I allocate any of those costs to the new product? The trouble is that if I keep producing new products without allocating any of my overheads to them then I will probably end up bankrupting the company. But if I allocate too many of my overheads then I won't win the contract to supply the toys. So I will come up with an amount it costs me to produce a cuddly polar bear and base my offer to the zoo on that figure, but it is important not to see that figure as some sort of scientific fact – it could have come anywhere on a wide range of costs.

In reality, I will probably take the additional costs of producing the polar bears and add a percentage to cover my existing overheads (for example, management, buildings and research costs that are not allocated to any particular product) to get a minimum price at which I can afford to sell. But there is another important concept here, which is sunk costs. My father used to give the example for sunk costs of the flower stall he used to walk past on his way home from work. Imagine this flower stall is only open on weekdays and does not have a way of keeping cut flowers fresh over the weekend. My father walked past this stall when it was about to close on a Friday afternoon and thought that he might buy my mother some flowers. Consider a bunch of roses that was being sold for £10. The stallholder paid £5. If Dad offered him £5 for the flowers it would make sense to accept it to avoid making a loss. What about if he offered £3? It feels like the stallholder should decline that because he would be making a loss, but in fact the £5 he paid for the flowers is a sunk cost – the money is gone and there is no way to get it back, so he should accept £3 for the roses. And nothing says romantic gift like economic theory and cut-price flowers.

In my cuddly-toy business, if the bottom suddenly falls out of the seal market and I'm left owning lots of white fur that I can't do anything with, I might offer to sell the polar bear toys more cheaply to limit my losses from the sunk costs, even if that means I have to sell the toys for less than it's costing me to make them. Selling them cheaply may mean that I lose less money than I would if I did not manufacture them at all.

Bidding for a polar bear contract is simple, but imagine if you're bidding for a contract to do something complicated such as building a bridge, when all of the costs are estimates because there are all sorts of unknowns. It may turn out that the bridge takes much longer to build than expected because of bad weather, or perhaps there will turn out to be an unexploded Second World War bomb found on the site. With so much uncertainty, it is very difficult to predict the costs of big projects accurately.

Take the London 2012 Olympic Games. When decisions were being taken about whether to bid to host the games, the people coming up with the costings had very little idea of what the eventual games would look like. The costs started at anywhere between £2.5 billion and £3.8 billion, but the games ended up costing closer to £9 billion. There is widespread support for the idea of evidence-based policy-making, but if the evidence is completely wrong, is it still a good idea? I asked someone who had been a senior civil servant at the time, and they said it was OK that the costings were wrong because everybody knew they were wrong. That is a very odd basis on which to make a decision. If everyone knows the costings are wrong then should the government just come out and say that the event will cost loads but it still thinks it would be a good idea?

The UK government did not face the same problems that the Swiss authorities had with trying to host the 2022 or 2026 Winter Olympics in the resorts of Davos and St Moritz. In each case, there had to be a referendum to decide whether to bid to host the games. I attended one of the events that was trying to persuade locals in Davos to support the idea and there was clearly a major charm offensive going on. In the end, local people rejected the idea of hosting the games in either year, with the way that costs always spiral from the initial estimates being cited as one of the key concerns among voters. I wonder if a referendum on whether to bid to host the London 2012 Olympics would have been successful.

The trouble with taking decisions about big, expensive projects is that you have to come up with a cash value for the benefits as well. In the case of the Olympics that involves a lot of tricky intangibles, such as inspiring the public to take more exercise. To take another example, the costing for the HS2 high-speed rail line between London and Birmingham and then Manchester and Leeds was particularly closely scrutinised, especially when it turned out that the benefits of it were based on the assumption that time spent on trains is completely lost time for businesspeople. Supporters of the line could claim that as it was going to cut half an hour off the journey time from London to Birmingham, it was reasonable to take the value of half an hour of the working time of all the business people using the service and add that to the benefits to the economy. But that is not really acceptable, because as long as you get a seat it is perfectly possible to do productive work on a train. Again, there is considerable guesswork involved in these figures. All big government spending decisions are based on such high-risk figures – does this make evidence-based policymaking a waste of time? I hope it doesn't but it is important to be aware of the uncertainty about any of the predicted costs or benefits.

How does any of this help you? You can avoid being misled by remembering to treat any claims involving cost with caution. Remember that costs for business and big projects are inexact. Once you have grasped the amount of room for manoeuvre with costings, you can think about incentives. Look at who is coming up with the estimate of how much something is going to cost and think about whether they would have any reason to want to make it look more expensive or less expensive. Would the headline writers like to have a bigger figure for the cost of snow or a smaller one? Would a government trying to launch a big project rather it looked more expensive or less expensive?

Once you have considered incentives and asked whether any of the costs would have been incurred anyway, you will be well on the way to grasping how much credence you should give to the costing in front of you.

CHAPTER 4

Percentages

Beware of lonely percentages

Percentages are a really useful tool, which can be used either to help or hinder people's understanding of numbers. Percentages are taught as part of the primary-school maths programme and I reckon most people you know will claim to understand them. But they are also the bit of maths I have been most surprised to find that colleagues have been unable to carry out.

Early in my career I worked for Reuters Financial Television, which made fairly pointy-headed economics and business programmes for City professionals to watch at their desks. I was programme editor for a show called *Equities Briefing*, which covered news about companies. It was a good, fun programme to make and colleagues were enthusiastic about working on it, especially after I introduced compulsory lunch breaks. One particular freelancer's job for the programme involved writing short pieces about company announcements, generally their results. We always included whether their profits before tax had gone up or down and the percentage change. Five minutes before the programme was due to start on her first day I looked at her script, which was excellent, except that it said the profits had gone up by x per cent – there was no number given. It was at this point that she admitted that despite having an economics-based degree, she could not work out a percentage change. Her memory of the incident is that I did not get cross (I'm glad to say) but instead we would work out the percentage changes together every morning as the director counted down to the programme.

She was not alone in struggling with percentages, but they are pretty simple really and very important in understanding the numbers around us. In this chapter I will go through:

- How to work out percentages
- How to tell if someone is trying to mislead you with percentages
- Compound interest and how it can help you understand big percentage changes

How to work out percentages

My colleague at Reuters went on to be very successful, I hope in part due to her newfound skill with percentages. Lots of people don't understand percentages, but most of them are not big enough to admit it. There are not many sums in this book, but when it comes to percentages there seems to be such a gap between the number of people who claim to be able to cope with them and those who actually can, that I am going to put the method here (nobody ever has to know whether you needed to read it). I see many colleagues using percentage calculators online now, and there's nothing wrong with that – we're not under exam conditions here. But there is something to be said for understanding how it's done from scratch – I think knowing the mechanics helps you develop a sense of when a number feels wrong.

There are three types of percentages you're likely to come across:

1. What is x per cent of this number?
   The comedy rock band Half Man Half Biscuit made a classic song called '99 Per Cent of Gargoyles Look Like Bob Todd', named after the actor who appeared alongside comedians such as Benny Hill. Milan's Duomo, which apparently has more statues than any other building in the

world, has 96 gargoyles (a distorted carved figure only counts as a 'gargoyle' if it is used to channel rainwater away from the building, otherwise it's a simply a grotesque). If you wanted to find out how many of the gargoyles on Milan's cathedral look like Bob Todd, and you accept that Half Man Half Biscuit's percentage is correct, you would start by working out 99/100, which is 0.99. Then multiply that by the number of gargoyles, 96, and you would discover that 95 of them look like Bob Todd.

2. What percentage is this number of that number?
   Returning to the Whiskas slogan that I mentioned in Chapter 2, if eight out of ten cats prefer it, what is that as a percentage? To work out the percentage, calculate 8/10, which is 0.8, and then multiply by 100 to get the answer: 80 per cent.

3. What is the percentage change in a figure?
   This is the one that catches out many people, certainly my assistant producer at Reuters Financial Television. There were 380 goals scored in the UEFA Champions League in 2016–17 and 401 goals scored in 2017–18. If you want to find out what that is as a percentage increase, subtract the old number from the new number (401 minus 380) to get the increase of 21 goals. Then divide that by the old number and multiply the

answer by 100 (this is where mistakes happen – people often divide by the new number). So that's 21/380 × 100, which is 5.5 per cent. So there has been a 5.5 per cent increase in the number of goals scored in the Champions League. If you want to try an exercise by yourself, there were 347 goals scored in the Champions League in 2015–16. Work out the percentage change between 2015–16 and 2016–17. The answer will be at the end of the chapter.

Staying on the football theme, people often get upset about players referring to themselves as having given 110 per cent, which annoys people because it is not possible. I'm usually at the front of the queue when it comes to numerical pedantry, but this one doesn't bother me so much because, firstly, we all know what they mean and, secondly, they could mean they are giving 110 per cent of the amount of effort they had given in the previous game, which is perfectly possible.

But it does introduce percentages over 100, which can be tricky. Take the headlines in 2017 about the cryptocurrency Bitcoin. In the first nine months of the year, the value of the currency rose from about $1000 to about $5000. What would you say the percentage change had been? You can go back to the method above and check before I tell you, if you like.

OK – it had gone up by 400 per cent. If you thought it was 500 per cent then you are not alone. Almost everybody on my statistics courses makes that mistake.

Remember, $1000 is 100 per cent of the original value.

$1000 to $2000 is a rise of $1000, so it's a rise of 100 per cent.

$3000 would be a rise of 200 per cent.

$4000 would be a rise of 300 per cent.

$5000 is a rise of 400 per cent.

What this tells us is that people just do not understand percentages over 100 per cent. What they do understand is multiples, so you're much better off saying the value of Bitcoin increased fivefold or by five times, and everyone will understand what you mean.

And before you dash out and buy a Bitcoin, bear in mind that the currency went on to a valuation above $19,000 in December 2017 before crashing to below $7000 in February 2018, so not one for widows and orphans.

Another pitfall to avoid when you're dealing with percentages is the difference between a percentage increase and a percentage point increase. This is important when you're comparing two percentages.

So, for example, imagine that you took an exam and you only received a mark of 25 per cent. You decided to take it again and this time you got 50 per cent. You could either say that your score in the exam has increased by 100 per cent (because it's doubled) or by 25 percentage points (from 25 per cent to 50 per cent). It's important not to confuse the two.

If you took the exam a third time and were given a mark of 55 per cent, how much better would you have done? Well, your score has gone up by five percentage points, from 50 per cent to 55 per cent. But you could also say that your score had gone up by 10 per cent, because five is 10 per cent of 50.

In November 2017, the Bank of England raised interest rates from 0.25 per cent to 0.50 per cent. This was widely reported as an increase of 0.25 per cent, which is wrong – it was an increase of 100 per cent (because it doubled) or 0.25 percentage points. Hardly anyone ran the headline 'interest rates double', as tempting as that might have been. You might also have heard it referred to as going up by 25 basis points or 25 bps. A basis point is financial jargon for 0.01 per cent.

Another conceptual difficulty with percentages is that they are not the same going down and coming up – if something falls 50 per cent and then rises 50 per cent it doesn't end up back where it started. If I buy a share that costs £1 and it falls 50 per cent then it will be worth 50p. If it then rises 50 per cent then it will be worth 75p. This is a particular hazard when you're following something that has fallen very sharply and then had a tiny recovery. It may have fallen so far that the tiny recovery is huge in percentage terms.

A good example of this is sales of vinyl records. Vinyl sales peaked in the 1970s when shipments reached about 90 million a year. This is shipments, not sales – if records were sent to shops and then not sold they would still appear on these figures. Also, it's the number of vinyl discs, so a double album would count as two. This might be a problem if the trend were ambiguous, but it's not – vinyl shipments fell from the late 1970s until the early 1990s and then collapsed to below one million. Falling from about 90 million to below one million a year is a drop of about 99 per cent, with first the CD and then downloads and streaming hitting the preferred medium of decades of rockers. The British Phonographic Industry (BPI), which represents the UK's recorded-music industry, started recording vinyl sales properly without the above caveats in 1994, charting their fall from about 1.5 million sales that year to about 200,000 in 2007. But then the recovery began, with sales passing one million again in 2014 and reaching 4.1 million in 2017. You could say that vinyl sales had fallen 99 per cent and then risen 1900 per cent, which would make it sound as if vinyl sales had gone through the roof and unprecedented numbers of records were being sold. Of course, that isn't the case. While the recovery in vinyl sales has been extraordinary, we are still very considerably below the level of 1970s peak vinyl, when 90 million discs were being shipped. Percentages are bigger on the way up than they are on the way down because nothing can fall by more than 100 per cent. You may be able to give 110 per cent of the effort you gave in last week's match, but you definitely can't give 110 per cent less than you did last time.

Now you know how to work out percentages and percentage changes, you know that you need to avoid percentages over 100, and you understand the difference between a percentage change and a percentage point change. You realise that percentages are bigger on the way up than on the way down, and you know how many of the gargoyles on Milan's Duomo look like Bob Todd. Armed with this knowledge, you can now start challenging some of the percentages reported in the news and get a feeling for when you might be being misled.

How to tell if someone is trying to mislead you with percentages

When you see figures only being given in percentages – what I call lonely percentages – you need to think about whether this has been done for a reason and whether the absolute numbers would tell you a different story. In the vinyl example, if you wanted to make it sound as if vinyl sales were back to their peak then you would just talk about the percentage increases. If you wanted to be clear that despite the extraordinary recovery we are still way below peak vinyl, then you would use the number of records sold.

A repeated argument during the EU referendum was whether the rest of the EU needed UK customers more than UK firms needed the business of the rest of the EU. This was used in arguments about whether or not the UK would get a good trade deal after Brexit.

Prominent Leave supporter Liam Fox talked a great deal about the EU's surplus in trade in goods with the UK – that means that other EU countries sell more of their stuff to the UK than the UK sells to the EU. The UK has a trading surplus in *services* with the EU, but it's not big enough to cancel out the deficit in goods.

Just looking at trade in goods, in 2015 the UK exported £134 billion worth of goods to the rest of the EU and imported £223 billion worth. So, if you were trying to make the argument that the EU needs the UK more then those are the figures that you would use.

On the other hand, in that same year, 47 per cent of the UK's total goods exports went to the rest of the EU, while only 16 per cent of the rest of the EU's exports of goods went to the UK. So if you were trying to argue that the UK needs the EU more then you would use the percentages.

If you heard trade figures being given in cash terms (billions of pounds) you were probably listening to a Leave supporter and if you heard them being referred to as percentages then you were listening to a Remain supporter. Both sets of figures are accurate, but they tell different stories.

While we're looking at the use of percentages in trade figures, Boris Johnson on several occasions described the UK as having been a relatively unsuccessful exporter in the Single Market. He said that in the 20 years after the creation of the Single Market in 1992, 27 other countries that were not in the EU had done better than the UK at exporting to the Single Market. The figure to which he was referring was the percentage increase in the exports of goods to the 11 founding members of the Single Market.

What he was talking about was not which countries had exported the most, it was which countries had achieved the greatest growth in exports. Top of the list was Vietnam, which saw its exports increase from $73 million a month to $400 million a month, a 544 per cent increase. That's impressive growth, but the actual cash amount of exports was still not huge at the end of the period.

The UK, on the other hand, was much lower down the list, having 'only' achieved growth of 81 per cent, but by the end of the period it was exporting $23.9 billion worth of goods per month to the Single Market.

Again, neither figure is inaccurate – it's not wholly unreasonable to define a successful exporter as one that has achieved growth. Indeed, growth seems to be what economists and politicians are most interested in. But without the cash figures as well you're not getting the complete picture. In this case, the UK is doing less well in percentage terms but much better in cash terms.

Percentages are extremely useful for putting figures into context, but if you're being given lonely percentages you should think about why this might be the case.

Another thing to look out for, if a figure does not sound to you like it is reasonably likely to be true, is whether the percentage has been worked out by dividing by the correct number. For example, the former Northern Ireland Secretary Owen Paterson told *BBC Breakfast* in December 2017 that the amount of trade between the UK and the

Republic of Ireland was 'quite small' and used some figures to support that claim. This immediately made me suspicious, because we know that countries being close to each other tends to mean they trade a lot. The Reality Check team looked into the figures that he gave, specifically that 5 per cent of Northern Ireland's exports go to the Republic of Ireland.

Mr Paterson pointed us towards the source of the figure, but it turned out that it was not 5 per cent of Northern Ireland's *exports* that went to the Irish Republic, it was 5 per cent of all goods and services produced in Northern Ireland that were exported to the Republic. He was dividing the amount exported to the Republic by the total amount produced in Northern Ireland, including things consumed in Northern Ireland. That is a strange way to measure exports.

If you work out the proportion of Northern Ireland's actual exports that go to the Republic the figure comes out at 37 per cent, which really cannot be described as 'quite small'. We were pleased to see that Mr Paterson changed his claim before addressing Parliament the following day, when he said 'only 5 per cent of Northern Ireland's sales cross the border south'. That is accurate but looking at exports as a proportion of a figure that includes domestic sales is somewhat unusual, and it was being used to support the claim that the problems of the Northern Ireland border after Brexit were 'easily surmountable'.

The problem in that example was not the figure for sales in the Irish Republic, it was what it was being divided by to come up with a percentage figure. If the percentage you are seeing feels a bit high or a bit low, check what is being divided by what. And while we're talking about confusion over what things are being divided by, just a quick note on another one to look out for.

The employment rate and the unemployment rate do not add up to 100 per cent. The employment rate is the proportion of the total population who are employed. The unemployment rate is the proportion of the economically active population who are unemployed. The difference between the two is the economically inactive part of the population, which is people who are not working and are not available for work or looking for work. I heard somebody on the radio talking about how there is a big gap in the employment rates for men and women of Pakistani origin in the UK because lots of Pakistani women wanted to stay at home and raise their families. This was statistically correct. Later, somebody on the same programme referred back to the research and said there was a gap between the two groups' unemployment rates, which is not accurate. Women staying at home to raise their families do not count as being economically active so they do not appear as part of the unemployment rate.

Another thing to look out for is when a figure hasn't been divided by anything but clearly should have been. I saw a headline on 22 May 2018 asking, 'Where are you most likely to be a victim of crime in Nottingham?' This was a local report (on the Nottinghamshire Live website) of Home Office figures on crimes such as burglaries, robberies and drug offences. The article said that there were 12,357 crimes recorded in the city centre, which makes it the most likely place to become a victim. But this takes no account of how many people live or work in the city centre or pass through it every day. You would expect several times more people to work in the city centre or visit it each day than you would see in a quiet, residential area. It surely can't be a surprise that there are more bicycle thefts in the city centre than there are in other parts of Nottingham. You can't possibly say where you're most likely to be a victim of crime without dividing by some measure of how many people there are in an area.

Percentages are often very useful, but be aware of what the indicator in which you're interested is being divided by to give a percentage rate, and don't be fooled by lonely percentages.

Compound interest and how it can help you understand big percentage changes

Compound interest is just jargon to describe how your savings add up faster if you leave the interest in the account. It's surprisingly important when you're trying to understand big percentage changes in the news.

Let's start with an example to illustrate how it works. To keep the sums easy, imagine you have managed to find a savings account that pays you 10 per cent interest – I know this seems unrealistic (imagine it's the 1990s or you're living in Madagascar or something). You have deposited £100 and you're leaving the money in the account.

Year 1: You get 10 per cent interest, which is £10, taking your total to £110

Year 2: You get 10 per cent of £110, which is £11, taking your total to £121

Year 3: You get 10 per cent of £121, which is £12.10, taking your total to £133.10.

By the end of Year 10 you will have £259.37.

By the end of Year 20 you have £672.75.

So you can see how the amount of interest you are getting each year grows because the amount of money in the account is growing. Over long periods of time the compound interest has a very powerful effect.

The best example of its use is in Douglas Adams's 1980 novel *The Restaurant at the End of the Universe*. The restaurant is enclosed in a vast time bubble and projected forward to the exact moment that the universe ends, allowing guests to eat sumptuous meals while watching all of creation exploding. The interesting element for our purposes is how you pay for it. 'All you have to do is deposit one penny in a savings account in your own era, and when you arrive at the End of Time the operation of compound interest means that the fabulous cost of your meal has been paid for.'

Time travel wreaks havoc with banking systems, but it's a nice idea. If you deposited your penny in a bank account today with a more realistic interest rate of 1 per cent, it would take you more than 200 years to get to 10p, and 463 years to get your first pound. But then it starts speeding up and after 1000 years you would have £209.59, which would be enough for a good meal for two, as long as you're ignoring inflation, which may put a spanner in the works. After 3240 years you would be sitting on £1 trillion. With the universe not expected to end for about six billion years, you can see how the money would add up, although you would have to be sure to pick a really stable economy and a bank with serious staying power.

Compound interest is not just about bank accounts. Oxfam released a report called *Growing a Better Future* in 2011 saying that global food prices would more than double by 2030. When you think about it, food prices operate in the same way as compound interest, because each year's increase comes on top of that of the previous year. The report predicted that food prices would go up by between 120 per cent and 180 per cent. The first thing to say is that that is a huge range, which indicates the massive uncertainty involved in predicting what is going to happen in 20 years. But is it a big number?

An increase of 120 per cent over 20 years is a rise of 4 per cent a year. An increase of 180 per cent over 20 years is 5.3 per cent a year. Is that a lot for global food prices? If you look at the food price index from the United Nations Food and Agriculture Organization (FAO), it turns out that food prices doubled in the ten years before this report was released. So actually, the predicted rises in food prices would have been a considerable slowdown in the rate at which food prices were rising, but that would not have made a terribly good headline. It was a shame in the end that the press release accompanying the report went with the line about food prices doubling, because it seemed to me that there were more important projections in the report. Knowing that prices are rising is not particularly helpful if you do not know what is happening to people's incomes. If everyone has much more money it does not terribly matter that prices are going up. Clearly this is not the case, however, and the more important prediction was that poorer people are going to have to spend a much higher proportion of their money on food than they do at the moment.

Hindsight is a wonderful thing and predicting the future is a mug's game, but if you look at the FAO's food price index now it turns out that 2011 was the peak for global food prices and they have been falling ever since. The index is based on the prices of five key commodity groups: meat, dairy, cereals, vegetable oils and sugar. The falls have been partly due to increased production, which might have been influenced by Oxfam's report, and partly because of the stronger US dollar – commodity prices tend to be listed in dollars so a stronger dollar means more can be bought for the same amount of money. From this we learn that if you can avoid making predictions for the next 20 years it's probably a good idea.

For the purposes of this chapter, the important thing is to understand that doubling over 20 years sounds like a much bigger rise than 4 per cent a year, but compound interest means it is the same thing. Making predictions over decades creates huge uncertainties, but if you understand what a big difference compound interest makes over time then you will be in a better position to understand the forecasts.

In conclusion, don't settle for the percentages and not the absolute figures. I will talk more about why this is important when discussing risk in Chapter 9 – it's very easy to scare people unnecessarily. Remember not to use percentages over 100 per cent because nobody understands what you mean. Hardly anyone gets the difference between percentages and percentage points, but you do now and the distinction is important.

Unlike many people you also now understand that compound interest accelerates your savings because it includes interest paid on the interest earned in previous years. So think about compound interest when you hear about big increases in something over a long period, or when you're trying to save up for a posh meal while the whole of creation explodes around you.

Percentages may be used to mislead people, but it won't work on you now that you know how they work. Oh yes, and the number of goals scored in the Champions League increased by 9.5 per cent between the 2015–16 and 2016–17 seasons.

CHAPTER 5

Averages

Know what you're talking about

The late lamented statistical genius Hans Rosling (1948–2017) pointed out that the average number of legs for people living in Sweden is below two. Nobody has more than two legs and a small number of people have fewer than two, so the average is very slightly below two. What this means is that almost everybody in Sweden (and indeed almost everybody in the world) has an above-average number of legs. The average does not in any way reflect the actual experience of Swedes.

This is an excellent example of an average that has been calculated correctly but is nonetheless unhelpful, because it ignores the way the numbers appear and says nothing about the real people in the data set.

This chapter is about averages and how they can be used to clarify a set of data or muddy the waters around it. We will learn how averages can be useful and representative, but that they are not necessarily an indication of the middle point of a range and may not reveal the whole story. Averages are a way to find a single figure that may tell you something useful about what is going on in the whole data set, but they may also conceal what is happening at the extremes or how much the numbers are spread out. Yet averages are used all the time on the news, generally without being challenged. There are many numbers in this chapter but don't be put off – none of the sums is difficult. This chapter will cover:

- How to work out mean, median and mode
- How you can be misled by the choice of average
- Measures of range

How to work out mean, median and mode

There are three measures of the average that you will come across: the mean, the median and the mode. The mean is what you get if you add up all the numbers and divide by the number of numbers. For example, if you're a cricketer and want to know your average score at the end of the season, you add up the number of runs you scored and divide by the number of times you batted. (Yes, cricket enthusiasts, I know that you need to take into account the number of times you were 'not out', but anyone who has ever seen me bat will know why that is not something that concerns me greatly.) When people refer to an average they are usually talking about the mean.

The median is the middle number. So, if you wanted the median test score in a class of 29 pupils, you would put them in order and take the fifteenth one. If there were 30 pupils you would take a figure half way between the fifteenth and the sixteenth.

The mode is a rarely used but occasionally useful figure – it's the number that appears the most often. If you wanted to know the mode age of professional footballers in France, it would be the age that the largest number of players are. If you took a survey of what method of transport people use to get to work, the mode would be a useful figure as it tells you the method that the largest number of people use.

For another good use of the mode, take a guess at what the average age of death was in England and Wales in 1964. The mean age of death was 65, so that's what you get if you add up the ages of all the people who died that year and divide by the number of people who died. But the mode is the age at which the largest number of people died. And in 1964 that was zero. More people died before their first birthday than at any other age. In 1964 that would have been no great surprise because it had been the case for most years before that. But it has not been the case since, and the fact that we now find it so shocking is a tribute to the extraordinary developments that have been made in healthcare, especially midwifery and neonatal intensive care. In 2016 in the UK, the mode age of death was 86. The mean was 78. And, if you're interested, the median was 81.

Why does this matter? Sometimes you will hear figures referred to as the average and you'll assume it's the mean when actually it's the median. It can make a big difference.

Let's take a look at an example using the ages of the starting team for Arsène Wenger's last home match as manager of Arsenal Football Club, against Burnley in May 2018.

Player	Age
Petr Čech	35
Héctor Bellerín	23
Calum Chambers	23
Konstantinos Mavropanos	20
Sead Kolašinac	24
Alex Iwobi	22
Granit Xhaka	25
Jack Wilshere	26
Henrikh Mkhitaryan	29
Alexandre Lacazette	26
Pierre-Emerick Aubameyang	28

Let's start by working out the mean. If you add up all those ages you get 281. Then you divide that by the number of players, 11, which gives you a mean age of 25 and a half. To work out the median we need to put all the ages in order: 20, 22, 23, 23, 24, 25, 26, 26, 28, 29, 35.

Then we take the middle number – the sixth one – because that is the player for whom half the team are older and half are younger. The median player in this case is Granit Xhaka, who is 25.

And there are two modes, which are 23 and 26 (as there are two players of each of those ages).

The mean and the median are pretty similar, with both of them broadly representative of the sorts of ages appearing in the Arsenal line-up that day.

Now imagine with the score at 5–0, Monsieur Wenger decided that in his 826th Premier League game he was sick of watching from the sideline and was going to bring himself on and show the youngsters how it's done. He warmed up, took his coat off to reveal a pristine red-and-white kit, and told the fourth official to hold up the board to replace the youngest player, Konstantinos Mavropanos.

Arsène was 68 years old, so let's see what he does to our averages. If you add up all the ages you now get 329. If you divide that by 11 you get a touch under 30, which is considerably more than the 25.5 mean we had previously. To find the median we have to write out the ages in order again: 22, 23, 23, 24, 25, 26, 26, 28, 29, 35, 68.

And the middle number – the sixth one – is now 26, one year more than before. The modes haven't changed.

We've added a number that is considerably different to all the others – what statisticians call an outlier. The mean has risen a great deal so that it's now higher than the ages of all but two of the players, but the median has only changed slightly.

And that's the point – you generally use the median if you want to prevent your average being skewed by any outliers.

Let's take another data set to be sure you're getting the hang of this. We'll go with the ages of the celebrity line-up for *Strictly Come Dancing* in 2017. I had great fun with this data set, checking for Reality Check whether it was possible to predict reliably at the halfway point which couples would get to the final. It turned out, not unreasonably, that the number of points the couples had gained so far was a pretty good indicator and that contestants tended to gain marginally fewer points from the samba, rumba, cha-cha-cha and jive. The most interesting thing was that while it looked as if there had been points inflation over the 14 series since the programme was launched, that was in fact a result of the series getting longer, which means the celebrities get more experienced and so were able to pick up higher scores. But the statisticians with whom I was working and I could not find any obvious biases in the system.

Here are the 15 celebrities together with their ages on the day the couples were announced, 9 September 2017.

Dancer	Age
Gemma Atkinson	32
Debbie McGee	58
Chizzy Akudolu	43
Ruth Langsford	57
Aston Merrygold	29
Richard Coles	55
Davood Ghadami	35
Simon Rimmer	54
Charlotte Hawkins	42
Mollie King	30
Alexandra Burke	29
Susan Calman	42
Joe McFadden	41
Jonnie Peacock	24
Brian Conley	56

Let's start with the mean. If you add up all the ages you get 627. If you divide them by the number of dancers – 15 – you get 41.8.

To work out the median you need to put the ages in order and take the middle one. The ages in order are: 24, 29, 29, 30, 32, 35, 41, 42, 42, 43, 54, 55, 56, 57, 58.

We need the eighth number, which is 42, so the mean and median are pretty close to each other. There are two modes, which are 29 and 42.

Now imagine that the world's oldest person was added to the line-up for *Strictly*. They are certainly a celebrity, but you would be surprised if they got very far in the show. Naming the world's oldest age-verified person at the time of writing would be foolhardy, so let's just take their age as 117. Now if you add up all the ages you get 744. If you divide that by the new number of contestants – 16 – you get the mean figure, which is 46.5.

To get the median you put the new age at the end of the above list of ages in order, and instead of taking the eighth figure you take a number halfway between the eighth and the ninth. In this case that doesn't make any difference and the median remains 42. There's also no change to the modes.

The introduction of a massive outlier in this case increases the mean by 4.7 years, but the median and mode remain unchanged. Once again, the mean has been shifted a long way by the outlier – 10 of the 16 contestants are now below the mean age. But the median has been unaffected.

In these examples we see how the choice of averages may influence how much your figures are affected by outliers. None of them is necessarily wrong as long as you have explained what you're doing, but you need to know which has been chosen, because there are great dangers of being misled.

How you can be misled by the choice of average

Now you know how to calculate the mean, median and mode, and we've started thinking about why you might choose one instead of another, we can start to look at some of the dangers of averages. How you choose to measure the average can completely change the impression given by a set of figures. Here are some of the potential pitfalls.

In the example at the start of the chapter, the mean was clearly the wrong sort of average to choose when trying to say something about how many legs people in Sweden have. The median and even the mode would have been better choices.

The choice of averages really does make a big difference. The Office for National Statistics works out average earnings across the economy every month based on the mean. For 2017 it was running at an average of about £500 a week. But the ONS also produces the Annual Survey of Hours and Earnings (ASHE), which has a figure for average weekly earnings calculated using the median, and for 2017 that came out at about £450 a week. Where has that £50 a week difference come from? It is almost certainly the case that a relatively small number of very high earners has skewed weekly pay across the whole economy by that much – it's about 10 per cent.

It's when you are looking at earnings that the differences between the different averages are particularly important. Consider the difference that Arsène Wenger made to the average age of the Arsenal team and now imagine how much difference it would make to average earnings if you had ten normal earners in the room and Bill Gates walked in.

Remember the package of tax cuts introduced in the USA by Donald Trump in 2017? The non-partisan Tax Policy Center calculated that taxes would be reduced by $1600 on average in 2018, increasing taxpayers' income after tax by 2.2 per cent on average.

Is that a helpful figure? It's going to look pretty inflated if you are one of the poorest 20 per cent of Americans, who will on average get a tax cut of $60, which is 0.4 per cent of their income after tax, or the next 20 per cent up who will get $380 or 1.2 per cent. In fact, the only group that sees an increase in its income of 2.2 per cent or more is the richest 20 per cent, whose average income rises by 2.9 per cent, which is worth an average $7640. The research found that 65 per cent of the benefit from the federal tax cuts would go to the richest 20 per cent.

While the average overall cut may be accurate, the vast majority of taxpayers will benefit much less, both in cash terms and as a percentage of their incomes. To see why that is, you need look no further than the benefit to the top 1 per cent of taxpayers who gain an average tax cut of 3.4 per cent, worth an average $51,140. The very large gains by the highest income Americans skew the average for taxpayers as a whole so much that at least 80 per cent of taxpayers are in income groups that can expect to see a smaller-than-average cut as a proportion of their income.

Income inequality means this is always something you need to look out for when you are considering averages across income groups. Relatively small numbers of big earners distort these sorts of figures a great deal if you're only using the mean. If there is no attempt made to explain how the money is distributed beyond a single figure then you probably aren't getting the full picture.

Another area where you may have seen different types of averages being used recently is in the figures for the gender pay gap. The first thing to remember is that whether a company has a gender pay gap and whether it has equal pay are not the same thing. The gender pay gap is the difference between the average amount men are paid per hour and the average amount women are paid. It takes no account of whether the men and women in the company are doing different jobs. Equal pay is about making sure that people doing the same or similar jobs are paid the same – it's a legal requirement. There are no reliable figures for how many companies are not offering equal pay because it is illegal. We have no idea what's happened to equal pay claims in employment tribunals because, since 2011, of the many thousands of equal pay claims that went to tribunals, 0 per cent were officially listed as successful at a hearing and 0 per cent were officially listed as unsuccessful at hearing. I'm assured by employment lawyers that this is because, on the whole, as soon as the legal points have been decided, the parties tend to settle the matter between themselves, but this is hardly justice being done in public, and does nothing to help our understanding of the scale of the problem.

Having a big gender pay gap does not necessarily mean that pay is unequal, but it may say something about hiring practices. A company running a chain of pre-schools, for example, would be much more likely to have women working in the nurseries than men, because almost all nursery teachers are women. That means that any men working in the business would be relatively more likely to work in management than looking after small children. This company may well have a big gender pay gap, despite offering equal pay to people doing the same job. On the other hand, a company with no gender pay gap could still be failing to offer equal pay. Imagine a company with four employees: a male and a female salesperson and a male and female manager. If the male salesperson was being paid a bit less than the female one and the male manager was being paid a bit more than his female colleague, there would be no gender pay gap but, all other things being equal, there might well be a case for unequal pay.

In the UK, companies employing 250 people or more have to tell the government their gender pay gaps, measured both by the mean and by the median. To get the mean gender pay gap you work out the mean average pay for the men in an organisation and for the women, and report the difference. It's the same with the median. Having both figures is useful – if your organisation has a much bigger mean pay gap than median pay gap it probably means that there are outliers, generally a small number of men being paid a lot.

But there are other tricky things to bear in mind when you are looking at gender pay gap statistics, in particular whether the figures are only for full-time employees or for all employees. The figures are reported as average pay per hour, so they would not be skewed by some people working more hours per week. The problem is that part-time work tends to be less well paid per hour and women are much more likely to do it. If you are trying to make your gender pay gap look smaller you might consider reporting only the gap for full-time staff.

It's these sorts of differences that meant that *Spectator* editor Fraser Nelson was able to say that, for women born after 1975, there is no gender pay gap – that's because he was only looking at full-time workers. If you're only looking at female workers, part-time workers get paid about a third less than full-time workers. And as women are much more likely to work part time, it means much of the gender pay gap is actually a part-time, full-time pay gap. Also, although there is still a gender pay gap, the situation has been improving, so only looking at women born after 1975 makes the figures look as good as possible.

If you were trying to make your gender pay gap look smaller you could also consider outsourcing some of the lower-paid functions in your company such as cleaning or call centres or catering. If you have more low-paid women doing these sorts of jobs, getting another company in to do them will remove them from your gender pay gap figures. The same would work if you were trying to reduce the amount more than the rest of the staff that the chief executive is being paid – outsource all the lower-paid functions in your company and suddenly your management is being paid a smaller multiple of the average salary.

When you're looking at these sorts of averages,

notice whether the organisation is using the mean or the median and check who is being included or excluded from the data. Also, think about what the data you are being told about is likely to look like. If you think there are going to be lots of extreme cases such as very rich people or very old people, you might question whether the average you are being given is really representative of the data set.

Measures of range

Michael Blastland and Andrew Dilnot's excellent book *The Tiger That Isn't* (2007) gives a great example of why averages may be misleading. A drunk swaying down the street between the two pavements will on average be walking in a straight line along the white lines. The traffic going in both directions will, on average, be able to avoid him. 'On average, he stays alive. In fact, he walks under a bus.'

If you're looking for somewhere to live with a comfortable climate you probably shouldn't look at the average temperature because you might end up in a desert that is very hot during the day and very cold at night, or somewhere that is extremely hot in summer and covered in snow all winter.

The problem with averages is that a single number may make it look as if everything is smooth and simple, when in fact that may not be the case. You would get the same average path for the drunk whether he or she was swaying between pavements or walking in a straight line. That's why in addition to being given the average figure, you may also be given a figure for the spread, or the range, or the deviation. Here's how it works.

There are various measures available for the spread of your data. In the case of the weather example, the simplest measure of the range would be to take just the maximum temperature and the minimum temperature. That would help if you were interested in avoiding extremes, but it would be less useful if, for example, you were looking at a place where the temperature was pretty much the same every day but there was one really hot day and one really cold day every year.

A better measure of spread, and one that you will see more often, is the standard deviation. The standard deviation is a measure of how far the figures are spread away from the mean – the lower the standard deviation the smaller the spread. If you were looking at the ages of a school football team taken from a single year group, the standard deviation would be close to zero.

Instead, let's work out the standard deviation for the ages of that Arsenal team against Burnley.

The first thing you need to do in order to work out the standard deviation is calculate how far away from the mean is each figure in your data set. Remember, the mean was 25.5, so we want to know how far each figure is away from 25.5.

Player	Age	Deviation from mean
Petr Čech	35	9.5
Héctor Bellerín	23	-2.5
Calum Chambers	23	-2.5
Konstantinos Mavropanos	20	-5.5
Sead Kolašinac	24	-1.5
Alex Iwobi	22	-3.5
Granit Xhaka	25	-0.5
Jack Wilshere	26	0.5
Henrikh Mkhitaryan	29	3.5
Alexandre Lacazette	26	0.5
Pierre-Emerick Aubameyang	28	2.5

There's no point taking an average of those deviations because that would give you zero. (It actually gives us a little more than that here because we've rounded the mean down to 25.5, but it's close enough.) Instead, we square all those deviations, so that the average will not be zero – we will take a square root later so it all works out. Squaring them has the advantage that all the figures we end up with are positive, which is good because we are interested in how far the figures are from the mean, not whether they are positive or negative.

Deviation	Squared
9.5	90.25
-2.5	6.25
-2.5	6.25
-5.5	30.25
-1.5	2.25
-3.5	12.25
-0.5	0.25
0.5	0.25
3.5	12.25
0.5	0.25
2.5	6.25

Now we take the mean of all those squares. They add up to 167. If you divide that by 11 you get a mean of 15.2. Finally, we take a square root of that number (because we squared them all earlier) and you get 3.9. That means the mean age of the team was 25.5 with a standard deviation of 3.9.

Now let's see what happens to the standard deviation after 68-year-old Arsène Wenger brings himself on in place of Mavropanos. If you want to work this one out by yourself now, grab a piece of paper. The answer will be at the end of the chapter.

While you're thinking about that one, let's sum up what we've learned. Averages are used to summarise a set of data at a glance, which is useful but sometimes dangerous. The mean, median and mode are each suitable for certain circumstances and misleading for others, and knowing which one is being used and why is crucial. A measure of how spread out the numbers are is also very helpful.

As a shortcut for situations when you may need to dig a little further into the figures, look out for qualifications such as the ones in the gender pay-gap example, when the average only covered full-time workers and those born since 1975.

A single figure to sum up a data set may be helpful, but it may also be misleading, and now you have the tools to spot which it is.

Now back to Mr Wenger's somewhat unorthodox substitution. With the manager playing on the team, the mean has risen to just under 30 and the standard deviation is 12.5, which is a big standard deviation, showing how spread out these figures are.

Just to illustrate the point further, we could go back to the starting team and replace the goalkeeper, who is himself an outlier in terms of age, with the substitute keeper David Ospina, who is 29. That would take the mean age down to 25 and the standard deviation to just 2.8.

CHAPTER 6

Big Numbers

Understanding billions, trillions and quadrillions

I was chatting with my 12-year-old son Isaac over breakfast on Budget Day. The children know it's Budget Day because I wear a hat, which says 'Hooray, it's Budget Day' that I had made specially. My wife tries to avoid being seen with me that day. There has been much discussion about what hat I should commission for other fiscal events: kudos to my colleague who suggested I should get one saying 'Public sector pay cap' and then I could raise it when necessary.

I was explaining to Isaac that Budget Day is when we find out how the government is going to be spending all the money. 'How much is the budget?' he asked me.

'About £850 million,' I sleepily told him.

'That doesn't sound like very much,' he said.

And of course, when I woke myself up and

thought about it, he was right. The government actually spends about £850 *billion* a year. He hasn't let me forget it since.

In this chapter I am going to talk about the challenges caused by big numbers. Really big numbers. Numbers so big it looks like somebody has fallen asleep with their nose resting on the zero key. Big numbers are difficult to deal with because our brains are not very well tuned to them. We can cope with small numbers because we have experienced them. You know what a group of ten people would look like and you could probably manage to visualise a crowd of a hundred. If you have been to a football game you might be able to cope with what 30,000 or even 100,000 people look like. If you have tried to buy a house you may have experience of hundreds of thousands, but there's very little in life that prepares us for dealing with billions or trillions. The very idea that £850 million could not be much money is one with which most of us struggle, even when properly awake. It's a growing challenge. When I first started working in business news in 1995, the total output of the UK economy was under £750 billion a year and that was about the biggest number with which I'd have to deal. The news contained millions and the occasional billion. Now, with the output of the economy approaching £2 trillion, there are more and more numbers in the news that sound big but are not. The government can announce that it's going to spend many millions of pounds on something, but in context that will be just a drop in the ocean. The classic example of this, cited in *The Tiger That Isn't*, was the pledge in the early years of Tony Blair's government to spend £300 million over five years to provide one million childcare places. So that's £300 per place, or £60 per place per year, which is not enough money for the policy.

There are three key techniques for making sure you're not being misled in this area:

- Double-check whether you mean millions, billions or quadrillions
- Put big numbers into context
- Remember a few key figures to help you understand big numbers

Double-check whether you mean millions, billions or quadrillions

Returning to my error over breakfast on Budget Day, I had made one of the most common errors in big numbers – saying millions when I meant billions. This will only become more of a problem as we report lots of trillions and even the occasional quadrillion.

A million is a one followed by 6 zeroes, a billion has 9 zeroes, a trillion has 12 zeroes and a quadrillion has 15 zeroes. If you have a nagging feeling that there is some difference between a British billion and a US billion, you used to be right. A British billion used to be a million million – that's 12 zeroes – while a US billion has 9 zeroes. Can you imagine the confusion that must have caused? In 1974, Robin Maxwell-Hyslop, MP for Tiverton, asked a written question of Prime Minister Harold Wilson, requesting that his ministers should agree only to use the British billion instead of the US version. The Prime Minister replied that a thousand million being a billion was now the internationally accepted version and his ministers would be using that. From that point, the British billion was abandoned and everywhere in the English-speaking world a billion is one followed by nine zeroes.

While I am confessing to my big number errors, I was once on *PM* on Radio 4 being interviewed by Eddie Mair, who has sadly now left the BBC. I was talking about trillions of things and he asked me how many zeroes that was. On the whole, live broadcasting does not make me particularly nervous, but suddenly at that moment I understood what happens to contestants on game shows under the studio lights when they give dumb answers to simple questions and we all throw stuff at the telly. I am eternally ashamed to say that I told him it was nine. He asked me another question – I started answering but my mind was elsewhere, and halfway through a sentence I stopped and said, 'No, hang on, a trillion is twelve zeroes.' The point is that anyone can make this mistake, so it's a good place to start when you're wondering if something is likely to be true.

Talking of mistakes, the first time I came across a quadrillion on the BBC News website was in October 2012 in a story about a woman in France who received a phone bill of just under 12 quadrillion euros. The exact figure was 11,721,000,000,000,000 euros. She phoned up her provider and suggested they might have made a mistake. They said it was definitely correct and offered to let her pay in instalments. In the piece, it was explained that even if she had paid in instalments equal to the entire output of the French economy it would still take her 6000 years to pay the bill. Now that's a big number. The phone company later admitted that it had made a mistake, said the bill should have been 117.21 euros, and waived it altogether, so there was a happy ending.

My only problem with the story was that it used 'qn' in the headline as an abbreviation for quadrillion. I know we use 'bn' for billion and 'tn' for trillion, but if you're going to use 'qn' for quadrillion it leaves you nowhere to go when we start reporting quintillions – that's a one with 18 zeroes. And indeed, we have had quintillions in the news. In 2014, when Psy's hit 'Gangnam Style' threatened to exceed the maximum number that the YouTube counters could cope with (just over two billion) the website updated its systems so that they could count up to 9.2 quintillion. I had thought that was the highest number ever reported by the BBC until a reader corrected me, pointing to a story from 2011 about a new system for allocating internet addresses, which created 340 undecillion possible addresses. An undecillion is a one followed by 36 zeroes, in case you weren't sure.

While these numbers may sound made up, they are not, unlike a zillion or a bajillion. But I often think that there is a place for made-up big numbers. They have been championed by US academic Stephen Chrisomalis, who calls them indefinite hyperbolic numbers. If we want to talk about, for example, single-use plastic straws, we know that loads of them are used each year, but we don't have a reliable estimate. So instead of coming up with an unreliable estimate that can be challenged, perhaps distracting people from the problem, we could say there are a squillion of them being used – it's a big number but we don't know what it is. This works well in conversation – 'I've told you umpteen times' is another good example. I'm not sure we're ready for it in more formal settings yet. 'The head of the NHS announced today that he would need an extra jillion pounds of funding over the next 10 years,' might sound a bit odd in the news headlines. It would certainly take some getting used to.

It is interesting, however, that there is a clear ranking of the size of these numbers, with umpteen being a relatively small non-specific number, a zillion or jillion certainly being more than a million, but not as much as a gazillion or bajillion, with the prefixes ga- or ba- increasing the perceived size of the number.

Put big numbers into context

To deal with genuine enormous numbers as opposed to made-up ones, you need some context, just as the French phone bill story used the economic output of France to get across what a ludicrously big number it was.

If your head is spinning with the size of all these numbers then you're not alone. My favourite attempt at helping came in a video called 'Obama Budget Cuts Visualization' on YouTube. It followed an announcement from President Obama that he was going to cut $100 million from the Federal Budget. We already know from my son that $100 million is not a lot of money when you're talking about a national budget, but maybe it's breakfast time and you need a bit of help with numbers this big. The video points out that taking $100 million out of a

$3.5 trillion budget sounds to the listener like taking a big pile of money out of a bigger pile of money. This is the key to the problem – when the numbers are this big we just can't visualise the difference between them. So, our filmmaker goes to the bank and gets 8880 one-cent pieces and puts them on his table in piles of five, each of which represents $2 billion – so each coin is $400 million. He picks up one of the pennies and cuts it in half with a pair of pliers. Then he takes one of the halves and cuts that in half again to get a quarter. He returns three quarters of the penny to the table and explains that President Obama plans to find a way to cut the remaining quarter of a penny from a Federal Budget represented by 8880 pennies. It's interesting that this visual explanation helps when other attempts have failed. The financial crisis was a terrible time for having to deal with big numbers – the figures involved were ludicrous. Sometimes, people would try to explain the bailout in terms of how much the money would weigh in £10 notes or how many times it would get to the moon if you stacked pound coins on top of each other. I think that was just replacing incomprehensibly big numbers with incomprehensibly big weights or distances.

The EU referendum campaign in 2016 threw up lots of interesting and contentious statistics, but one above all others has become notorious and is the enduring image of the campaign. I am talking, of course, about the £350-million bus. The Vote Leave campaign wrote on the side of its campaign bus: 'We send the EU £350m a week – let's fund our NHS instead'. The problem with it was that we did not send that amount of money as our contribution to the EU Budget because the rebate was deducted before any money was sent. The rebate is a discount that the UK gets on its contributions to the EU Budget – it was originally negotiated by Margaret Thatcher in 1984. BBC Reality Check pointed out the problem with the figure when it was being used in interviews, long before it appeared on the bus. Later, the UK Statistics Authority (UKSA), which is the UK's independent statistics regulator, ruled that it was potentially misleading, so any further discussion was just arguing with the referee, which is pointless. The UKSA was particularly worried about the idea that the £350 million could all be spent on the NHS, because part of it is the rebate, so is not spent at all, and parts of it are spent by the EU on things in the UK such as supporting farmers, scientific research and regional aid, all of which would leave people cross if they thought their funding was being diverted to the NHS. Even the most enthusiastic users of the figure now refer to the £350 million as the amount of money over which the UK loses control, rather than the amount the UK sends to Brussels. The £350 million was not in itself a misleading statistic – it is the correct figure for the UK's contribution to the EU Budget before the rebate is taken off, but that's not how it was described on the bus.

The £350 million presented two separate big-number problems. The first one was that the correct figure for the amount that the UK was sending to the EU each week was £276 million. Clearly there is a difference between £276 million and £350 million – it's about 25 per cent – but to anyone listening it was just two big piles of money. And every time somebody argued that the bus should say £276 million instead of £350 million, it still highlighted the fact that the UK was contributing a big pile of money to the EU Budget.

The second big-number problem was the context in which £350 million, or indeed £276 million, was not really a lot of money at all. The Institute for Fiscal Studies and the International Monetary Fund both pointed out that the UK's contribution to the EU Budget would be completely dwarfed by any impact on the economy as a whole. The output of the UK economy is about £2 trillion a year. If Brexit means that the economy grows by 1 per cent a year more than it would have done, or by 1 per cent less, the difference it would make to government finances would be more than the contribution to the EU Budget.

We have seen how figures can be used to put other figures into context, making big numbers more manageable and easier to understand. But not all numbers would work with this. If you divided the size of the US Federal Budget by the number of cats in the USA, you would get a smaller number expressed in terms of spending per cat, but that wouldn't be very helpful.

The question is, which numbers do you use to put figures into context, and that depends very much on the sort of number with which you're dealing. If you're talking about funding for education it may be worth dividing by the number of pupils or the number of schools. In the example at the start of the book, dividing the number of plastic straws used per year by the population meant you could immediately take a view on whether the figure is likely to be true. If you can personalise the number and relate it to something you can comprehend then you will probably find it easier to cope with.

For a large sum of money, a good option is to compare it with areas of government spending. Once you start talking about amounts of money at national budget levels, there are always going to be big-number problems. And it's not just whole government spending that creates these difficulties – the National Health Service is so big that it's hard to talk about anything to do with it without losing people in the numbers. The budget for NHS England is about £115 billion a year, which is an unimaginably large amount of money. I suggested at the 2015 general election that we should get round the big-number problem by doing what physicists do. A light year is the distance that light travels in a year, which is about 5.9 trillion miles or 9.5 trillion kilometres. That means you can express mind-bogglingly big distances as just a small number of light years, making it easier to cope. If we take an NHS yearly budget as £115 billion then other large amounts of money could be expressed in terms of how long they would fund the NHS. An NHS month would be just under £10 billion and an NHS week would be just over £2 billion. The £350 million on the bus would have been a touch more than an NHS day. The slight problem with the NHS year as the yardstick for large amounts of money is that it is not a constant – it changes from year to year, whereas a light year is always the same distance. On the other hand, that means it incorporates a measure of inflation, which may be useful when considering large amounts of money. In other words, as things get more expensive across the economy, so too does an NHS year.

The NHS is also a big employer with about 1.5 million workers, which makes it the world's fifth biggest employer behind the US Department of Defence, the Chinese army, Walmart and McDonald's. Such scale makes some news about the NHS a bit difficult to assess. Take the time an anti-cuts campaign group released the shock news that the NHS was planning to cut 53,000 jobs. If you read in a bit further it turned out that the group had put in freedom of information requests to calculate the number, which was spread over the next five years. I'm not taking a view on how much the NHS would be poorer for having 53,000 fewer staff, or indeed the damage that would be caused to those households by the loss of employment. My problem with the analysis was that I didn't believe it was accurate enough to be correct to within 10,600 jobs a year out of a total of 1.5 million. So 53,000 jobs sounds like a lot, but that's because all headlines about the NHS sound like a lot, be they reports of unfilled posts, extra funding or budget deficits. Try to think about such reports in the context of what an extraordinary institution the NHS is.

Another useful example came from a headline in the *Daily Mail* on 2 September 2009: 'Town hall bans staff from using Facebook after they waste 572 hours in ONE month' (that's the *Mail*'s capitalisation). It turned out the story was about Portsmouth City Council, which was blocking access to Facebook for its 4500 staff. What are you going to divide by to establish whether this is a big number? Start by converting 572 hours into minutes to make the sums easier – it's 34,320. Then divide that by the 4500 staff to get 7.6 minutes. And that's 7.6 minutes a month spent on social media by each member of staff – if you divide again by the 21 working days in a month you get to about 22 seconds a day per employee, which really isn't a big number. Later in the story, it turns out that some members of staff are supposed to be using social media to check up on the lifestyles of benefits claimants. So really the headline should have been: 'Town hall massively overreacts to tiny use of social media by staff'. That's also a good story, perhaps even a better story than the original. It illustrates another point, which is that numbers in headlines are enormously powerful. Almost everybody reading that article would have come away thinking that Portsmouth City Council employees were wasting huge amounts of taxpayer-funded time posting things on Facebook, which just wasn't true, but there is no question that is what the *Mail* wanted you to believe.

Although context is important when you're dealing with big numbers, some people get upset about the ways that big numbers are put into context.

When I first started as the BBC's head of statistics I went to give a talk at the Office for National Statistics. At the end, I asked the audience what annoyed them most about numbers in the news. One of the delegates asked me if I could persuade the BBC to stop referring to areas as multiples of the size of Wales (the ONS headquarters is in Newport). I said I couldn't do that because coping with big numbers in the news relies on things being the size of Wales, the volume of an Olympic-sized swimming pool, enough to fill Wembley Stadium, the length of a jumbo jet or the height of a number of double-decker buses.

Clearly there are problems with this. I objected to the reporting of a chimney due for demolition as being the height of 55 double-decker buses stacked on top of each other on the grounds that if you stacked that many buses on top of each other the bottom ones would get squashed. It was much more sensible to describe it as more than double the height of Big Ben (or the tower containing Big Ben, although that doesn't trip off the tongue in the same way). I suspect people have little idea how big Wales is, but saying something is the size of Wales probably means more to them than saying it's just over 8000 square miles or almost 21,000 square kilometres.

I have also seen people try to use the size of an elephant as a standard measure for something big. I received research suggesting that the amount of reusable rubbish being sent to landfill weighed the same as 90,000 elephants. That feels like it's just using one big number to replace another and not really helping with the context at all, not to mention the fact that there is a big range in the size of elephants. Perhaps they would have been better off using the weight of Wembley Stadium or indeed the weight of whales – or even Wales.

The cash comparison that I find unhelpful is comparing people's salaries to that of the Prime

Minister. We're always hearing how the chief executives of some councils or NHS trusts are paid more than the PM. The PM's salary is kept artificially low for political reasons, at about £150,000, but that doesn't include the rent-free residences that come with the job: 10 Downing Street and Chequers. The use of an attractive Central London residence and a mansion in the country must be worth a fair bit.

Similarly, you often hear rich people or valuable companies being described as worth more than particular countries. This is a bogus comparison because the value of the country is almost always measured by its gross domestic product or GDP, which is the amount its economy produces in a year. That is then compared with the total wealth of an individual or the market capitalisation (the value of all the shares put together) of the company. Clearly this is not comparing like with like. If you compared an individual's salary or a company's sales for a year with a country's GDP it might make more sense, although it's still not terribly illuminating.

We have seen how dividing a giant phone bill by the economic output of France or the amount of time spent on social media by the number of staff in a town hall has helped with our understanding of big numbers, while dividing the US Federal Budget into 8880 on YouTube demonstrated how $100 million wasn't really a big number. It's not tricky maths – it's just using division to provide the context that helps you understand whether or not you should be getting excited about particular numbers.

Remember a few key figures to help you understand big numbers

It's useful to have a few key figures at your fingertips to help you understand big numbers. It's fine if these are rounded up or down because the bar is set at the level of 'reasonably likely to be true'. I'm as bad as anybody at remembering or estimating some of these. The ONS put together a quiz in which you had to say how many people lived in your local area and how many of them had jobs or university degrees, for example. I was terrible at it. I suspect most people were. The pollsters Ipsos Mori carry out an annual poll in 40 countries called 'The Perils of Perception', in which it turns out that we have very little idea about things such as what proportion of the population come from ethnic minorities, what proportion are homeowners and how much the country spends on healthcare.

Here are ten statistics about the UK, rounded to make them easier to remember (or perhaps you could bookmark this page if you can't remember them). These figures are all from the ONS.

- UK population is about 65 million – 55 million in England, 5 million in Scotland, 3 million in Wales and 2 million in Northern Ireland; almost 9 million live in London.
- About half of those people are employed; about three-quarters of those aged 16 to 64 are employed.
- There are about three-quarters of a million live births in the UK a year and 600,000 deaths.
- The total output of the UK economy measured by GDP is about £2 trillion a year.
- According to the 2011 Census, 86 per cent of the population of England and Wales is white; the next biggest group is Asian/Asian British at 7.5 per cent followed by Black/African/Caribbean/ Black British at 3.3 per cent.
- 59 per cent of the population of England and Wales identify as Christian, 25 per cent have no religion and 5 per cent are Muslims; the Census question was optional and 7 per cent didn't answer.
- Just over nine million of the UK population were not born in the UK while about six million are not UK nationals.
- About 65 per cent of UK households are owner-occupiers, 17 per cent rent from a private landlord and 18 per cent from a social landlord.

- The average (median) weekly wage for a full-time employee in the UK is £550; that's £28,600 a year.
- The UK's national debt is about £1.7 trillion.

Challenging big numbers is all about confidence. If you have a few carefully chosen numbers at your fingertips, you can use them to put the figures you hear into context. Also, you can be pretty sure that your friends and family don't know the ten indicators above off the top of their heads, so it's a great opportunity to get ahead in arguments.

Do bear in mind that everybody confuses millions and billions occasionally so you always need to double-check. When you see a figure in a headline that looks big, stop to think about whether it really is, especially if it involves government spending or debt, which always throw up huge sums.

You're now ready to deal with big numbers.

CHAPTER 7

Correlation and Causation

Did this really cause that?

I heard a frustrating interview on the radio, introduced by the newsline: 'People admitted to hospital with a head injury are twice as likely to die over the next 13 years as people who haven't suffered that kind of injury.'

There was then an interview with the chap from the University of Glasgow who had led the research. We learned from him that the people who had suffered the head injuries were dying of just the same things as the rest of the population and it was 'not entirely clear' why their death rate was higher. They had tried adjusting for gender, age and social deprivation, but that didn't explain away the effect.

The professor said he was now going to do some more research and find out if there were other lifestyle factors involved (in other words, are people who end up in hospital with head injuries more likely to do things that get themselves killed within the next 13 years?) or whether there is some other 'underlying biological cause'.

The possibility that was not mentioned was that this was all just a coincidence.

This chapter is about correlation – two things are correlated if one goes up or down at the same time as the other. It does not mean that they have anything to do with each other. We hear all the time in the news about how eating particular things makes us more or less likely to get cancer, for example. It's really hard to prove that one thing is causing another and it is certainly not safe to assume it is just because both have gone up at the same time.

Correlations are very useful if you are looking for things to investigate, but they can create problems in the news when they are used without further evidence to suggest that one thing is causing another.

In this chapter we will look at the three questions to ask when you assess things that are correlated:

- Is this a coincidence?
- What else is going on?
- Are the numbers oddly specific?

Is this a coincidence?

Modern spreadsheets are so powerful that you can put pretty much anything going on along one axis and pretty much everything else along the other and find out if anything correlates to anything else. Remember, a correlation is when one thing goes up or down at the same time as another thing goes up or down. It doesn't mean that one thing is causing the other.

This is beautifully demonstrated by a website called Spurious Correlations, which plots charts showing correlations between things such as the amount of cheese eaten per person in the USA and the number of people killed by becoming tangled in their bedsheets. There was also a clear correlation between the number of people who drowned after falling into a pool each year with the number of films Nicolas Cage appeared in.

In *The Tiger That Isn't*, Michael Blastland and Andrew Dilnot give the example of somebody throwing a bowl of rice in the air. When the rice ends up on the ground, there are some places with little piles of rice and there are other areas where there is no rice at all. And we all just understand that this is the way it is with rice falling on the ground. But when we are talking about things that sound like they might be linked, such as people dying in the years after they have suffered a head injury or unusually large numbers of people in an area getting a particular type of cancer, our brains are attuned to looking for patterns, and the explanation that this is all just a coincidence is desperately unsatisfying. Blastland and Dilnot offer the explanation that this is a survival instinct – it's better to be safe than sorry when deciding whether the pattern in the trees is just an illusion caused by light and shifting leaves or a real tiger – hence the title of their book. It's hard to fight the urge to find patterns and things causing other things when we see sets of numbers, but we need to try if we are going to avoid being misled. This is all metaphorical, incidentally. If there's a chance there might be a real tiger involved then my advice is to be on the safe side and run for it.

When the radio presenter said that the head-injury case was 'quite an alarming finding', she was right, but it shouldn't have been. It's certainly an interesting finding, and a good basis for applying for research funding to investigate whether there is something going on, but until there is any reason to believe that it's not a coincidence it shouldn't really be getting out to an audience that could be unnecessarily alarmed. After all, it's not as if people will be going around trying to get head injuries if they don't think it will increase their chances of dying of something else in the following years.

Demonstrating that two things that correlate are also causing each other is very hard. In health research, the way to show something causes something else is with a randomised controlled trial (RCT). In a classic RCT you get a bunch of people and randomly allocate them into two groups, one of which is given the treatment you are testing and one of which is given a placebo (something like a tablet with no active ingredient), and neither group knows who is getting the new treatment. Because the allocation is random, there is a decent chance that any differences in the outcome for members of the groups will be due to the treatment you are testing.

Now consider how you would run an RCT to show that being admitted to hospital with a head injury caused people to double their chances of death in the following 13 years. You would take a group of people and divide them randomly into two groups. Then you would beat all the members of one group about the head hard enough for them to need to be admitted to hospital. After that, you would wait 13 years to see how many members of each group had died. It would be tricky to get this piece of research past the ethics committee, which is why people try to use other statistical methods.

So, to confirm, it is very difficult to demonstrate that one thing is causing another without either using an RCT or finding the mechanism by which it happens.

The great exception that proves this rule is smoking. It is now generally accepted that smoking causes cancer, although there have been no RCTs, because the weight of evidence is so overwhelming even without a full understanding of the mechanism. Again, you would struggle to run an RCT on smoking because you would have to find a group of people and randomly choose half of them to start smoking and the other half not to, and then observe who got cancer in the following years. The causal link being accepted between smoking and cancer is unusual.

Another big advantage of RCTs is that you decide in advance what you're looking for, making the outcome much less likely to be a coincidence. Well-designed experiments also mean you can't just do the test over and over again until you get a significant result. As a simple example of this, it's very unlikely that you will flip a coin ten times in a row and get heads each time – the odds are about one in 1000. The illusionist Derren Brown once did this on television using a single continuous camera shot, but he later explained that the footage that was broadcast was just the last minute of nine hours of filming him failing to get ten heads in a row. It was in a programme called *The System* (2008) in which he managed to send five consecutive correct predictions about which horses would win a series of races. It turned out that the way he had done it was to take a group of thousands of people, divide it into six and send the name of one of the six horses in a particular race to each group. Then in the next round he would take the group of people who had been sent the winner of the previous race, divide that up into six and so on until the only person left was one woman who had received five consecutive correct predictions. The woman, of course, believed that Derren had the unfaltering ability to predict the outcome of any race. It was an extraordinary lesson in how we can be led by our own perceptions and how badly we can be misled. He used it to make a point about things such as homeopathic remedies. If we take something when we're feeling unwell and we end up feeling a bit better, we will be absolutely convinced it is what we have taken that has made us feel better regardless of whether tests on thousands of people have found them to be ineffective. We have taken something at the same time as we have started feeling better so we assume that one thing has caused the other.

In his 1954 classic *How to Lie with Statistics*, Darrell Huff talks about the use of vaccines and antihistamines as cures for the common cold. As a cold will eventually cure itself, given time, you can give people any sort of treatment you like and be reasonably confident that they will feel better within a week or so. You can then give the credit to whatever treatment you have recommended and many people will be convinced – personal experience is extremely powerful.

When it comes to designing research experiments, as well as planning the experiment in advance it is also good to have a hypothesis that you are testing. Some newspapers and magazines were taken in by a hoax piece of research, conducted in 2015 by a journalist called John Bohannon, which fooled them into reporting that eating chocolate accelerated weight loss. Fifteen volunteers were recruited and divided into three groups – one group was put onto a low-carbohydrate diet, one was put on the same diet but told to eat a 1.5oz bar of dark chocolate every day and the third group, the control group, was told to carry on eating as usual. At the end, the control group had not lost any weight at all. The two groups on diets lost an average of five pounds in three weeks, but the group eating the dark chocolate lost the weight a bit faster.

This sounds like a well-designed study and it was published in a journal and in some newspapers. What was not included in the reports was the fact that the study examined 18 different measurements such as weight, cholesterol, blood protein levels and quality of sleep. In a study of only 15 people there was likely to be a false positive, caused simply by coincidence, in at least one of the 18 measurements. In this case, it turned out to have been speed of weight loss, so that was what was reported.

There are a few problems with this story that should have been spotted such as the absence of references online to 'Johannes Bohannon, PhD', the person cited as being behind the research, or to the Institute of Diet and Health that he set up for the project. Also, while there is considerable detail in the paper about the people involved in the study, it does not say how many of them there were, which is a fairly basic piece of information. On the other hand, there has been some criticism of the way John Bohannon decided to try to fool members of the public in order to show that it is easy to fool members of the public. I hope that people deliberately setting out to mislead people in this way are rare. And it is a salutary lesson in the importance of checking sample sizes and ignoring diet research that has only involved small numbers of people.

One of my favourite suggested correlations came with the news that the head of NHS England had decided to ban the sale of super-sized chocolate bars in hospitals. A spokesperson for the biggest operators of shops in hospitals was quoted in the papers saying that by introducing such schemes early, sales of sushi and salad had been boosted by 55 per cent, while fruit sales were up one quarter. The implication there is that customers looked at the range of snacks available, noticed there were no 'King Size' chocolate bars, and decided to buy sushi instead, which is a difficult scenario to imagine. Actually, it turns out that they no longer make King Size bars – they are often called 'Duo' now, so people can pretend they are going to share it with a friend – like 'Sharing Bags' of crisps. Following further enquiries it turned out that what the hospital retailers had actually said was that they had introduced a whole 'Healthier Choices' programme and a range of ways to encourage people to eat healthier snacks, so it wasn't just the banning of enormous chocolate bars that had done it.

There is a cartoon I love from the webcomic *xkcd* in which one person says: 'I used to think correlation implied causation. Then I took a statistics class. Now I don't.' His friend says: 'Sounds like the class helped.'

'Well, maybe,' the first speaker replies.

What else is going on?

It's important to point out that the problem with confusing correlation and cause is not just about missing the possibility that the two things are a coincidence. It may also be that there is something else going on. Radio 4's fabulous *More or Less* team mocked up a story about how mobile-phone masts were increasing the birth rate. They had found a correlation between the number of masts in an area and the number of babies born. It turned out that for every extra mobile-phone mast in an area, there were 17.6 more babies born per year than the national average. Is there just something romantic about masts protruding from the countryside?, the programme asked. There is no question that the two things are correlated and it's not a coincidence either. Telecoms companies tend to put more mobile-phone masts in areas where lots of people live so they can get a mobile signal. There are also more births in areas where lots of people live, for reasons that should not need explaining. So there is a link, but there is an extra step that needs to be taken before it makes any sense.

It may be that the correlation you are looking at is just a coincidence. That's a good assumption with which to start. But the next question to ask is whether, as is the case with the mobile-phone masts, there is something else going on – another step to take.

I had a letter from my son's school talking about its poor attendance record (the school as a whole, not just my son). The letter said: 'While we would never claim a direct causal link, statistics suggest that improving your attendance by 1 per cent could lead to a 5–6 per cent improvement in your attainment.'

I am certainly glad that they did not claim a direct causal link, but the wording of it and similar statements from the Department for Education is certainly meant to imply that one thing may be leading to the other. It is particularly tricky because the letter was complaining about the high levels of applications for planned absence. It does not appear to distinguish between approved absences and children just not turning up on a particular day for whatever reason. It seems likely that there is another step in this process. Having the sort of stable family life that means you are being told to get out of bed and go to school every day is likely to be causing other things that are good for your educational attainment such as being encouraged to do homework. While having a great deal of unauthorised absence would suggest you might not be getting this sort of support at home, that is not the case with authorised absence. The Department for Education's statistics show that pupils eligible for free school meals – generally taken as the indicator of whether pupils are from low-income families – are on average absent from school more. At Key Stage 2 (although not for other age groups) it turned out that pupils who had more authorised absence achieved a moderately better than average level of attainment, although again I would caution against any suggestion that there was a causal link. So the question to ask yourself is whether the reason some pupils achieve lower levels of attainment if they miss a few days of school is because there were crucial lessons being taught on those days or because there might have been some other reason why they missed school such as ill health or a chaotic home life that is also lowering their attainment. My son's school may have a poor attendance record, but it doesn't have a poor academic record.

Another aspect of the question about what else is going on is whether you're considering a causal link the wrong way round. The government is looking into whether there should be limits to the amount that teenagers use smartphones. It has cited an increase in anxiety among girls as justification for this. There have been studies suggesting that teenagers who spend excessive time on their phones are more likely to be anxious or depressed, but you could just as well argue that teenagers who are anxious or depressed are more likely to spend excessive time on their phones. The main trouble with this is that almost all teenagers use mobile phones, so while more depressed teenagers are likely to use phones, so are ones who are not depressed. You particularly need to look out for this factor before blaming something on universal activities. For example, most children are vaccinated, so most children who end up with various ailments have had those vaccines. Just demonstrating the correlation is not enough – many children who have been vaccinated do not suffer from those ailments. Without an RCT or an understanding of the mechanism by which one thing causes another, you have to assume that it's a coincidence.

Are the numbers oddly specific?

An aspect of the story about people with head injuries that should make you wonder what is going on is the fact that the period being considered was the 13 years after somebody had been in hospital with a head injury. Why would you be interested in the following 13 years? If you were choosing parameters before starting the research you would surely choose 10 or 15 years. The choice of 13 years suggests that the time period may have been chosen to give the most extreme results. That's not the only possibility, of course – it may be that data is only available for the last 13 years and the researchers were keen to use as much of it as possible.

But it is easy to look at a chart of something over time, pick the highest point and the lowest point and claim a trend, while ignoring what happened before or afterwards.

Look out for periods such as 13 years, which you could describe as 'oddly specific'; this was also the name of a now defunct website that was full of examples of figures that were surprisingly precise. There was a picture on it with an American-style road sign saying that the speed limit was 15¾. I'm sure it had been Photoshopped because there was a sign behind it that said 'Beware of signs', but it makes the point. Sometimes you expect a round number and if you don't get one you should wonder why.

While I was thinking about this subject I went for a walk round the block and it turned out that my local estate agent had a big sign up with a picture of a bottle of champagne saying 'Celebrating 28 years. Please ask us about our special celebratory offers.' I had never thought of the twenty-eighth anniversary as a particularly important one. I found a website suggesting that in parts of southern Europe it's the amber anniversary and that you could celebrate it with the gift of an orchid. But what the sign makes me think is that the estate agent is not cutting its profit margin to celebrate its significant birthday but that it's just trying to get you through the door. I know, I'm a desperately cynical man and I should take joy in helping my local estate agent celebrate, but the oddly specific number put me right off.

The sense that these things sound funny clearly kicks in early in life. My six-year-old dashed into the kitchen the other day, laughing himself silly, to inform me that he was watching 'The Simpsons' 138th Episode Spectacular'.

Here's another one: there was an advert on the London Underground for a dating website. It claimed that the site had led to 144,000 Britons being in relationships, adding 'that's 2,208 Tube carriages of people!' I wonder why 2208, given that means there are 65.2 people in each carriage. Would fewer people have signed up for the service if they had thought it had only got 2200 carriages full of people into relationships? I am a happily married man and would not have been signing up either way, but presumably there are people out there who are as pedantic about numbers as I am, but still single – I know it's hard to believe.

Oddly specific numbers are sometimes jarring because they do things with figures that are too precise for the context. This is particularly the case for forecasts – if somebody tells you that something will cost the economy a certain amount over the next decade to the nearest pound, you know they are overstretching the accuracy of their model (leaving aside the bogus nature of all such predictions, which we discussed earlier).

There was a report in 2018 that suggested pro-Leave orchestrated bots on Twitter could have created a mismatch between Leave and Remain content on Twitter that translated into 1.76 percentage points of difference in the actual voting. The same thing in the US presidential election helped Donald Trump by 3.23 percentage points. The mechanism between content being shared on Twitter and people going out and voting seems dubious at best, but to suggest that it can be cited to the nearest 0.01 of a percentage point feels as if it's pushing the point considerably further than is justified.

A wine grower came on Radio 4 during the heat-wave summer of 2018 and said that English wine producers were exporting to 'over 26' countries worldwide. If that's making you wonder what the actual number was then you are getting the hang of this. If it is 27 then why not say 27? If it is 28 then why not say it was over 27? If you're using 'over' then the number really needs to be a round one, otherwise it is oddly specific.

You can hear yourself when it sounds strange. If you were saying that something in France had cost you about 100 euros you wouldn't convert that to £88.15 because the number you were starting with wasn't that precise. You would say it had been about 100 euros, which is about £90.

The main things to remember when you're being told that something is causing something else are that your first assumption needs to be that it's a coincidence, and if there is a link, you need to ask what else is going on. Has immigration increased youth unemployment or was there a recession going on at the time that might have had something to do with it? Be particularly careful when the causal link you are examining is something you really want to believe – something that matches your existing beliefs, such as the reports about chocolate accelerating weight loss. This is also why you need to be careful with research conducted by charities. Charities do excellent work, but sometimes the statistics they release are a bit weak. Our brains are already programmed to look for patterns and assign causes, so if on top of that we want to believe what is being said because we like the organisation, then we need to be on our guard. After reading this chapter, I hope you are ready to be wary of correlation being dressed as causation. That may be a result of the advice the chapter contains or it may just be a coincidence.

CHAPTER 8

Alarm-bell Phrases

What to look out for

There are various words and phrases that should set off alarm bells in your head every time you see or hear them. They don't necessarily mean that the figures you are about to hear are wrong or misleading, but they make it more likely. You will see them in adverts and the news, you will hear them from politicians and you should take them as a hint that you may be being misled. Understanding the numbers properly is all about seeing the warning signs at the right time, because you can then choose to ignore a story completely or seek more information. Some of these phrases come up regularly, such as 'sickie' statistics every time there is a major sporting event taking place. Others may be hidden in important speeches, where you wouldn't notice them if you were not on your guard. Once you start listening out for certain phrases, you'll find them all over the place.

Up to

'Up to' is a widely used and generally misleading weasel phrase, which means that the figure with which you are being presented is the maximum possible. Those signs on the high street saying 'Always up to 60 per cent less' seem to me a copper-bottomed guarantee that the shop will never discount anything by more than 60 per cent. It does not mean that anything in the shop is actually reduced by 60 per cent. In fact, it doesn't necessarily mean that anything in the shop has been reduced at all, just that nothing has been cut by more than 60 per cent.

Similarly, I saw a sign in a department store saying 'Our gift to you. Up to 30 per cent off selected lines'. You won't be surprised to hear that '30 per cent off' was in about 500 point, bold lettering, while 'up to' and 'selected lines' were in about 10 point. But the 'selected lines' disclaimer is interesting, because presumably it could mean that non-selected lines will be more than 30 per cent off.

We know that this is the sort of thing that retailers do and we have become used to it, but when it gets into the news it's a problem, especially when it spoils one of my favourite stories. *Metro* reported in December 2012 that South Yorkshire Police had spent £7000 on 280 cardboard cutouts of officers in an attempt to deter thieves. 'The cardboard officers have reduced crime by up to 50 per cent in some areas,' the paper said. First of all, I'm dying to know how South Yorkshire Police demonstrated that any fall in crime was due to the presence of cardboard officers and not other factors. But secondly, we don't know from that report that crime has been reduced at all, only that crime has not been reduced by more than 50 per cent in some areas. And it follows that it has been reduced by more than 50 per cent in other areas.

More seriously, the Conservative Party manifesto in 2017 promised to recruit 'up to 10,000 more mental health professionals', a promise that the government can keep as long as it does not recruit more than 10,000. At the same election, the Liberal Democrats pledged to: 'create a new designation of national nature parks to protect up to one million acres of accessible green space'. When making election promises it seems to me that a minimum number would be more appropriate than a maximum.

There are many other similarly unhelpful expressions, including 'as much as', which means the same thing as 'up to', and 'at least', which is the opposite. If you are not interested in the maximum or minimum levels for a particular figure, then don't use them.

As if the regular use of 'up to' wasn't bad enough, it's often used in ways that make it even worse. Buzzfeed ran a headline saying: 'As many as 300 managers at BBC News earn up to £77,000 or more, according to a leaked document.' To break down what that means, there are no more than 300 managers who are earning either more than £77,000 or less than £77,000. That's not a newsworthy revelation.

'Sickie' statistics

You would not believe the number of times you see statistics used with great confidence to trumpet the number of people doing something they shouldn't. It is remarkably difficult to collect such figures because people tend to go out of their way to prevent anyone finding out.

I often receive emails from PR firms claiming to have discovered how many people are planning to 'take a sickie' on a particular day, usually when there is an event such as an important football game happening in the middle of a weekday. These figures are based on surveys. I look forward to the day that someone phones me to ask whether I'm planning to call in sick so I can watch England playing on the television. Are there really people who say yes? The survey results I'm sent suggest there must be. The follow-up to the number of people planning to take an unjustified sick-day is always how much that will be costing the economy, but you already know from Chapter 3 that such claims are dubious.

The sick-day story is frivolous and (I hope) mainly harmless, but the same principle goes for figures about illegal activities. Some statistics on law-breaking come from reports or surveys of the victims, which is fair enough although still not an exact science. But take a hefty pinch of salt with figures such as illegal migration, illegal downloads or using mobile phones while driving.

Record numbers

We all love a record. When someone wins Olympic gold, breaking the record is a cherry on the top of their achievement and you will find me cheering as much as anyone. But most aspects of life are not like the Olympics. You achieve your record age every second. Populations tend to grow and prices tend to rise, which means that you would regularly expect there to be record numbers of people doing things and that record amounts will be spent on things, both by governments and individuals. Prime Minister Theresa May regularly talks about how record amounts of money are going into education, which may be a surprise for people with school-age children who keep hearing how short of cash their local schools are. While the amount of money in the pot for schools in England is indeed at a record high, rising pupil numbers and prices mean that schools are going to have to make savings of approximately 8 per cent of their budgets by 2020. Record spending is not necessarily good enough because the things that schools spend money on would be expected to get more expensive every year: teachers, buildings, heating, lighting, books and stationery are all likely to cost more. And there was a baby boom in the early 2000s, which is working its way through the school system. What you need to know is what has happened to spending per pupil after you have adjusted for rising prices (inflation).

Death tolls

When you hear figures for the number of people killed by a disaster, think about how those statistics have been collected. I remember seeing pictures in the days after the Boxing Day tsunami in 2004, in which whole towns had been washed away, and yet the official death tolls were still only about 10,000 – the eventual figure was more than 200,000. Anybody in the area trying to help has better things to do than try to come up with an accurate figure for the numbers of people who have died.

Remember the controversy about the number of people who died in the appalling Grenfell Tower fire of 14 June 2017 in London? These were official police death-toll figures, which were announced as 12 people, then later 30 people, gradually rising to the final figure of 71. The police went to extraordinary lengths to make sure they were certain before announcing somebody had died, searching every flat in the building for human remains, which was made very difficult by the damage caused by the ferocity of the fire. When you hear official police figures early on, they are likely to be conservative, for perfectly good reasons.

There is considerable uncertainty about death tolls in war zones. The figures for the numbers of people killed in Syria vary wildly, while the best estimates for the numbers killed in Yemen are based on figures from hospitals, even though many of the hospitals have closed as a result of the conflict and much of the fighting happened in rural areas that never had hospitals.

This isn't just a problem for death-toll figures – any statistics are hard to collect in conflict zones. It is difficult to find a representative sample for a survey in a developed country that is at peace. Imagine how much more difficult it would be to do so in a war zone.

International comparisons

Headline writers love being able to compare statistics between countries, but it's a dangerous thing to do. Take the headlines in March 2015 claiming that people in Rwanda can expect to live more years in good health than the most deprived 10 per cent of people in England. The figures for Rwanda came from the World Health Organization (WHO), which found that people born in the country in 2012 had a healthy life expectancy of 55 years, which was a remarkable improvement, given that in 2000 the figure was 40.

Journalists then compared that with figures from the Office for National Statistics (ONS), which looked at the inequality in healthy life expectancies between rich and poor people in England. It found that the most deprived 10 per cent of the population had a healthy life expectancy of 52 years, while the least deprived 10 per cent could expect 71 healthy years. That's a shocking range, but does the comparison with Rwanda work?

To check, you can look at the WHO research. It also gave a figure for the UK, which it said had a healthy life expectancy of 71, the same figure that the ONS was giving for the best-off 10 per cent in England.

This should make you wonder if the two studies, while using the same terminology, used different methods. And that was indeed the case. The WHO figures were based on a detailed survey about people's health and used the answers to subtract years of ill-health from existing life-expectancy figures. The ONS used the results from its Annual Population Survey, which asked respondents if their health was very good, good, fair, bad or very bad. It turned out that people were more likely to classify themselves as being in ill-health under the ONS system than they were to be classified as being in ill-health based on the answers they gave to the WHO. That means that the comparison between deprived people in England and everyone in Rwanda does not stand up. The WHO also warned about the difficulties of conducting such research in low-income countries, which also makes such international comparisons difficult.

If you really need to make international comparisons, see if you can get a big, global organisation to do it for you: the World Bank, the United Nations and suchlike are good at this sort of thing. Their websites are less user-hostile than they used to be and their statistics divisions are very helpful.

Straw poll

When a poll or survey is dubious or inaccurate people try to justify this by saying it's just a straw poll or a snapshot. What that means is that the methodology is not good enough but they are going to try to get away with it anyway. It's 'just a bit of fun' in Peter Snow's words. But if the results are being used to try to convince the public that opinion is in one direction or another then it's serious. I am told the term 'straw poll' comes from the idea of holding up a piece of straw to establish which way the wind is blowing, which seems like a more robust methodology than you see from straw polls.

Remember the morning after the general election in 2017, many outlets reported that there had been 72 per cent turnout among 18-to-24-year-olds? This was the start of the 'youthquake', which ended up being the *Oxford Dictionary* word of the year. The figure was tweeted by Labour MP David Lammy and also by National Union of Students president Malia Bouattia, who said she was not surprised. I was certainly surprised given that the pollsters at Ipsos Mori estimated that only 43 per cent of that age group had voted in 2010 and 44 per cent in 2015. The figure was eventually tracked down to Alex Cairns, who ran an organisation trying to encourage young people to vote – he stressed that the figure was 'an indication' based on conversations he had conducted with student union presidents and other research that he had done. So we can add 'an indication' to the list of alarm-bell phrases.

We never find out exactly how many people from any particular age group vote in a general election because it's a secret ballot, but the best indication we get is from the British Election Study, which concluded in January 2018 that there had been little change in youth turnout between 2015 and 2017, with the margin of error meaning that it could have risen a bit or fallen a bit.

These words and phrases should put you on your guard, but they do not mean that what you are reading is definitely nonsense. One straw poll that was entirely robust came from Pig World, the voice of the British pig industry. It was wondering how much rising straw prices were hitting the margins of pig farmers, so it told its reporter at the British Pig and Poultry Fair to go and ask some of them what they thought. The reporter said she had spoken to farmers who were being squeezed by the price rise. There was no attempt to use these conversations to come up with any questionable stats and the publication ran with the headline: 'Straw poll highlights concerns over a straw-based future'. Excellent work.

Statistically significant

Statistically significant is a technical term, which means that the findings are outside the margin of error. You need to be careful not to confuse that with something being significant, meaning important. By convention, if a finding in a piece of research is statistically significant it means that if that same research were to be conducted 20 times, that finding would turn up at least 19 times. Remember that the more people you ask in your survey or the more patients you have in your study, the more likely your findings are to be statistically significant because the margin of error gets smaller. A really huge study will probably find lots of things that are statistically significant, but that doesn't mean they matter.

In a speech in 2017, Royal Statistical Society president Sir David Spiegelhalter gave the example of research suggesting that watching television for more than five hours a night increased your chances of having a fatal pulmonary embolism, compared with watching less than two and a half hours. But he added that when you looked at the absolute level of risk, it 'could be translated as meaning that you can expect to watch more than five hours of television a night for 12,000 years before experiencing the event, which somewhat lessens the impact'. So the effect may have been statistically significant, but it may not have been very relevant to people's lives. This is not to say that watching five hours of television a night is not bad for you in other ways – just that you can expect other problems to manifest themselves considerably earlier than the blocked arteries in the lungs.

Prof. Spiegelhalter also warns against assuming that if doing an enormous amount of something is bad for you, doing a little bit of it will also be. Look out for that logic in health-related articles.

Beware of the word 'significant' with any qualifier attached such as 'borderline significant', 'potentially significant' or even 'practically significant'. A one-in-

20 chance of your findings being a coincidence is not setting the bar particularly high. If the research can't even get over that bar then it's almost certainly not good enough. Also, 19 out of 20, or 95 per cent confidence, is a pretty standard place to set the bar. If you read something that is trying to set it at 90 per cent or lower you should be wondering why.

People asking not to be named

Anonymous sources have their place. If a journalist is interviewing victims of crime or even perpetrators of crime it's generally pretty obvious why people would want to remain anonymous. But if it's a story littered with anonymous sources and it's harder to work out why they would want their names withheld, you should be on your guard. One of the wire services used to have a reporter covering energy prices who came up with colourful quotes from unnamed oil traders such as: 'the blood is in the water and the sharks are getting frenzied'. They used to brighten my days, but I did wonder why he was having to quote unnamed traders. Did he have a friend who was coming up with these lines but wasn't really allowed to talk to the press or was he just making them up himself?

Exclusive statistics

'Exclusive' is an overused word in news that makes me wonder who is being excluded and why. Sometimes it's obvious why an article is called an exclusive – some hard-working journalist has managed to find something out or get access to something or someone, and has used this to come up with a story that nobody else has. But when you hear that a news organisation has been given exclusive access to a particular data set, you should probably wonder why. Working in a busy newsroom, you become attuned to the signs that a press release is going to contain unreliable figures, and one of the key indicators is that the organisation sending it to you is prepared to offer you exclusive access to it. Sometimes the lure of an exclusive story is enough to make journalists let down their guard and run some questionable figures of which they would otherwise have disposed. Again, this does not necessarily mean that a statistic is nonsense, but it should put you on your guard.

I don't have to look far back in my inbox to find 'new research' (I hardly ever get sent old research) being offered to me as an exclusive. Top of the list is the revelation that 'over two-thirds (64 per cent) of Generation Z say salary is their top motivator in the workplace'. What a great story, even once you get past the belief that 64 per cent is more than two-thirds. I decided against publishing a BBC exclusive revealing that members of Generation Z like being paid to work.

Television viewing figures

TV viewing figures are another one of those statistics that sound like they are a precise count of a total number of people, but are in fact based on a survey. Five thousand households with 12,000 individuals log their viewing habits and the figures are then extrapolated to cover the whole of the population. Especially with so many channels available, it means that one or two people changing their viewing habits may have an enormous impact on the ratings for a particular programme. This is not the case for online traffic though. On my second day working for the BBC News website, I received an email telling me precisely how many times the article I had written the day before had been clicked on. It was terribly exciting, especially having come from television where no such precision exists.

Even though the Barb television rating system used in the UK is based on a survey, it's a pretty big survey. The figures from it are certainly more robust than the figures you see now and then for the number of people watching particular sporting events around the world. I have seen claims that the Tour de France is watched by 4 billion, 3.5 billion and 1.6 billion people.

Another figure to be deeply suspicious of is how many records a particular artist has sold around the world, which are widely reported, usually when they die. These global figures simply weren't collected, especially in the early careers of some of the ageing rockers we find ourselves mourning. The figures are desperately unreliable and may be considered estimates at best.

Comparisons over a long time

Early in the 2016 EU referendum campaign, Michael Gove claimed that the proportion of UK trade done with the EU was lower than it had been before the UK joined the EU. I ended up in London's Guildhall Library, working my way through the five volumes of the 1972 edition of the *Annual Statement of the Overseas Trade of the United Kingdom*. Then over the weekend I realised I had ignored trade in services and would have to dig up a separate publication called *Britain's Invisible Exports*. At that point a friendly economic historian took pity on me and helped me find the right figures on a United Nations database, and I could establish that the proportion of trade with the EU had not fallen as claimed.

This brought home to me how spoiled I am when I try to find out how much a statistic has changed since last year, or over five years, when I can just look online. Some departments at the ONS in Newport still proudly display their huge, leather-bound volumes of historic statistics, but I don't generally have to use hard copies.

There are problems with making comparisons over long periods of time. In the same way as indicators with the same name differ from country to country, so the methodology for indicators in one country may change over time. There may be warnings that series are not continuous when there have been big changes, but it is still easy to be caught out. For example, the Census for England and Wales, which is conducted every ten years, asks how many rooms a household has. In 1971, respondents were told not to include kitchens that were less than six feet wide. In 1981 and 1991, kitchens less than six feet six inches were ignored. In 2001 and 2011, all kitchens counted as rooms, so if you look at the whole series, it will look as if rooms per household increased more than they actually did, because more small kitchens were being included.

Two of my favourite recent claims including very long time series are that pay growth was at its slowest since the Napoleonic wars and that governments since 2010 had borrowed more money than all Labour governments put together. Both of them were fun to check because I got to talk to economic historians and use the Bank of England's excellent 'Millennium of Macroeconomic Data' research data set. Nonetheless, earnings figures from the 1800s are sketchy at best and trying to compare borrowing by the Ramsay

MacDonald government in 1924 with borrowing by the coalition government in 2010 ignores all sorts of changes to the country and its economy.

When you see claims that something is the worst it has been for hundreds of years, take them with a suitably large pinch of salt.

Official statistics

We hear many references to official statistics, but what does that mean? The gold standard of statistics in the UK are national statistics – they are easy to spot because they have a kitemark with a tick, saying 'national statistics' on it. The rules governing how they are produced and presented are strict. Every now and then a national statistic loses its kitemark because of concerns about its methodology. The point is, national statistics are tightly policed – while they are not perfect they are probably the best-available figures. Government departments also produce what are called 'official statistics', which are still governed by the Code of Practice for statistics but might not be quite as robust or as important. Other figures released by government departments are not necessarily as reliable, and journalists who are short of time may accidentally describe them as official just because they have come from the government. The trouble with these other releases is that ministers may get involved in which figures are placed most prominently and there may not be consistency across publications. You should be as suspicious of a set of figures from a government department that does not classify as an official statistic as you would be about numbers from any other organisation.

Poverty statistics

When you see figures for poverty they will either be measuring absolute poverty or relative poverty. Relative poverty tends to mean that a household has an income after taxes and benefits that is less than 60 per cent of the median income. Remember, the median income is the one that half of households are above and half of households are below. These figures are generally also adjusted for the size of the household because a bigger household needs more to live on. There are advantages to this measure – it tells you whether a household is able to afford things that other households can. But it throws up odd situations, such as if the economy is doing really badly then the median income is likely to fall, which means that a household may suddenly be lifted out of relative poverty, not because it's better off but because lots of other households are worse off.

Absolute poverty is generally considered to be a situation in which someone cannot afford basic necessities such as food, clothing and shelter, but that's not how the UK government sees it. Its measure of absolute poverty is the same as the measure of relative poverty except that instead of using the median income at the time, it uses the median income from 2010–11. This makes things a bit confusing. All I can suggest is that when you hear about changes in the number of people living in poverty, make sure you know what measure people are talking about.

You should be wary of figures for global inequality. In particular, some of the highest-profile measures of wealth inequality allow for negative wealth, which means that the poorest people are not those living with almost nothing in slums, they are people with enormous mortgages living in huge houses, or those with considerable student debt who might nonetheless have high-paying jobs. These are not generally the people you think of as being at the bottom of the global wealth pile.

Underlying

I don't want to talk too much here about company results, but there are a few key alarm-bell words to listen out for when chief executives tell you how their companies have been doing, and they are 'adjusted' and 'underlying'. The rules for preparing financial results are designed to use a number of fairly arbitrary assumptions to make one company's results broadly comparable with another's. But company bosses like to claim that the message you are getting from the usual headline figures such as pre-tax profits or net earnings is not really telling you what's going on in the company, so they produce 'adjusted' figures, which are supposed to tell you about the 'underlying' performance of the business.

Sometimes this is fair enough. If a company has finally decided to write off all the money it spent on a misguided takeover five years ago, which means that it has to report an overall loss of many billions of pounds, it may also be useful to have a figure excluding that write-off so that you can see how the rest of the business has been doing.

On the other hand, sometimes you get the feeling that the adjustments are not being made to help you understand the business, and the more things that are excluded or adjusted for, the more likely it is that you are not seeing the full picture.

There was an American company providing serviced office space, which tried to focus on a figure it called community-adjusted EBITDA. EBITDA is a fairly standard earnings figure, excluding factors such as interest payments, tax and the falling value of things such as buildings and equipment (it stands for 'earnings before interest, taxes, depreciation and amortisation'). But the community-adjusted bit meant that they were also excluding the amount they were paying their employees, advertising expenses and the costs of opening up in new locations. The community-adjusted EBITDA figure was positive, suggesting the business was profitable, unlike the overall earnings figure, which showed a net loss of almost $900 million.

Maybe the company is right, and its huge profitability is being masked by various items that do not really affect the strength of the business. But it certainly suggested to me that further investigation was needed.

Once you start listening out for certain words and phrases you will hear them all the time. They are an excellent way of reminding you that you might be being misled and should either look a bit further into the claim being made, or ignore it and get on with your day. In particular, now you know to be wary of phrases such as 'up to', 'straw poll' and 'adjusted', you can start thinking about how some statistics try to quantify things that are difficult to ascertain. It can mean that both the statistics you are presented with and any others with which they are being compared may be skewed.

CHAPTER 9

Risk and Uncertainty

How likely is that not to be true?

In February 2014 there was confusion about the UK unemployment figures, which meant that some news organisations were reporting that the unemployment rate had fallen to 7.2 per cent and some were reporting that it had risen to 7.2 per cent. It had actually fallen, despite the previous month's figure having been lower. You may think this is unreasonably confusing and you would be right.

This chapter is about uncertainty, risk and chance, which are hugely important parts of statistics but get less attention than they deserve. When you see big statistical claims being made strongly in the headlines, you should consider carefully whether they are justified. Many of the statistical methods behind those claims will not be good enough to provide anything more than an estimate, but the words used will suggest unjustified precision. The flip side of asking how likely something is to be true is asking how likely it is not to be true, which is all about getting across the uncertainty involved. When we talk about how likely things are to happen, we are talking about risk; understanding and communicating levels of risk is one of the great challenges facing statisticians, journalists and politicians. Nobody has really cracked it yet, but if you understand where some people are getting it wrong and how others are trying to get it right, you will be in a much better position to know what is going on.

The three tools you need in this area are:

- Using measures of uncertainty to find out if a story is really a story
- Making sure you see both absolute risk and percentage changes
- Deciding whether the odds of more than one thing happening are being considered correctly

Using measures of uncertainty to find out if a story is really a story

When you see the unemployment figures published each month, you may think it's a precise count of the number of people who would like to have a job and do not have one, but it's not – it's based on a survey. That's a perfectly normal way of measuring unemployment, and the current UK unemployment statistics are the best-available figures. But it's important to understand what the methodology does to how precise the figures are.

Almost all of the UK's official statistics are based on taking a sample and assuming the rest of the country has broadly the same features as the chosen few. In fact, the ONS baby-names statistics are among the small number of figures that are based on counting everyone involved with no extrapolation. That means you can tell exactly how many babies were called a particular name in each year, as long as there were three or more of them. I can tell you that there were precisely 6259 baby boys named Oliver in England and Wales in 2017, and 3 named Oliver-John.

Most of the more important UK statistics, though, are an estimate and it is important to use the correct language about them so as not to sound too precise. Let's start thinking about the uncertainty involved by going back and looking at why unemployment figures have been unreasonably confusing. To do that, we need to look at how the figures are collected.

Unemployment statistics

The headline figure for unemployment used to be the number of people claiming unemployment benefits, which is called the claimant count, but then it was realised that the government could fiddle the numbers by moving people onto different benefits such as disability benefits. The new Universal Credit system makes the claimant count even harder to assess in terms of unemployment, because some Universal Credit claimants are employed and some are unemployed. As a result, the claimant count has lost its status as a national statistic. To get away from just counting the number of claimants, for many years the ONS has used an internationally agreed definition of unemployment as its headline figure: somebody is unemployed if they do not have a job, have tried to find a job in the last four weeks and is available to start work in the next two weeks.

To work out that figure, the ONS conducts a Labour Force Survey (LFS), a huge survey in which it talks to about 40,000 households containing 100,000 individuals every three months. It starts with staff going and knocking on people's doors all over the country, with follow-up calls coming from a call centre at the ONS offices in Titchfield, on the south coast of England. You may have seen a documentary series on BBC Three called *The Call Centre*, which followed the activities of Nev Wilshire, the boss of a Swansea call centre, and his staff of mainly 20-somethings as they battle to get out of bed and get anybody on the phone to talk to them. Well, the

ONS call centre in Titchfield is nothing at all like that. For a start, it's mainly staffed by older women who have been working there for years. I visited a few years ago and listened into some calls. It's completely unlike the sort of cold-calling you usually hear from people conducting surveys. The people they call are expecting to be rung and seem to have a relationship with the system.

The point is, this is a seriously high-quality survey, but it's still a survey, which means that there is a margin of error. The ONS is very good at telling us what the margin of error is for all of the measures in the labour market report. It does that with the use of confidence intervals, which show the range of values within which it thinks the actual figure is likely to be. So, for example, the figure for the change in unemployment generally has a 95 per cent confidence interval of about 75,000, which means the ONS is 95 per cent confident that the actual change in the number of people unemployed is within plus or minus 75,000 of the figure they have given. If the change in unemployment is smaller than that confidence interval, the figure is described as not being statistically significant, which means we can't say that unemployment has either risen or fallen. If you look at all the figures for 2017, there is not a single month when the change in unemployment is significant. Any headline you saw that year based on quarterly figures that said UK unemployment had risen or fallen was misusing the statistics.

Another thing that makes these figures tricky is that they are quarterly statistics – they cover a three-month period – but they are released every month. So the January to March figures will be published in May and then the February to April figures will be published in June, and so on. That's because the people who ask the questions from the Titchfield call centre have the households they are surveying divided equally over the 13 weeks of the quarter. It means that two-thirds of the data used each month was also used the previous month. Each month's figures will be based on one-third of the respondents having been asked since the last figures were released and two-thirds of respondents for whom they are still using the answers they gave in the previous two months. As a result, you shouldn't be comparing this month's figures with last month's figures – you should be comparing them with the ones that were released three months ago.

This brings us back to the situation described at the beginning of the chapter. The ONS reported that the unemployment rate for October to December 2013 was 7.2 per cent. Some news outlets reported that as an unexpected rise, because they were comparing it with the previous month's figure (covering September to November) of 7.1 per cent. But the figure was comparable not with the previous month's figure but the one for the previous quarter, July to September, which was 7.6 per cent.

The system of reporting rolling quarters every month does make life difficult for journalists trying to report the figures. Up until the late 1990s, unemployment was reported once a quarter, but then it was realised that the figures could be released monthly without having to collect any extra data. There are ways round the problem. One of them would be to triple the number of people interviewed in the Labour Force Survey, but that would be eye-wateringly expensive. If the government was thinking of injecting a bit of extra cash into the collection of employment statistics, it's also worth bearing in mind that even doubling the number of households surveyed would only make the change in unemployment correct to plus or minus about 55,000 instead of about 75,000. Another option would be to headline the change in unemployment compared with a year earlier instead of the previous quarter, which does give you lots more significant changes but also makes the statistics feel less current. Or, as a drastic measure, news organisations could consider only reporting the unemployment figures when there has been a statistically significant change, but I can't see that happening.

High-profile statistics

Unemployment is one of the most closely watched statistics published each month. Other figures that always get lots of attention are inflation, migration and GDP, all of which involve some sampling.

The inflation figure tells us how much prices are rising. It is based on a notional basket of goods and is the figure targeted by the Bank of England's interest rate-setting Monetary Policy Committee. The statistical authorities decide what is a normal range of items to buy in a month and then look at how the prices of those things have changed in a range of outlets. It's a sample of the changes in prices across the economy. If you buy a drastically different range of products to the ones the statisticians have chosen, or indeed use different suppliers, then you will experience a different rate of inflation.

The migration figures are based on asking a selection of travellers at ports whether they are planning to come to the country or leave the country for at least a year. If you think that's an inaccurate way of working out migration, and indeed the population, then you're not alone. Its accuracy is sensitive to factors such as people changing their plans, refusing to be interviewed, lying or arriving at ports where there are not ONS staff asking questions. For example, many of the passengers arriving from eastern Europe in the mid-2000s were coming to smaller, regional airports, where the International Passenger Survey had no representatives. Just as with unemployment, the sampling method means that the rise or fall in immigration has to be pretty large to be statistically significant, so we can't always say with confidence whether migration has risen or fallen.

The early estimates of GDP, perhaps the most influential economic statistic, rely heavily on extrapolating what's happened to the economy based on actual results from a relatively small proportion of companies. By the time the figures have had their fourth or fifth estimate and are based on more actual returns from businesses and fewer estimates, it's probably too late to change everyone's perception of what is going on in the economy, and may be too late to change any measures taken by governments or central banks based on what they had been told was going on.

To be fair, the ONS is in the process of carrying out all sorts of interesting research into how it can use other sources of data to improve the accuracy and speed of its statistics. You could use records from GP surgeries to improve population data or scrape prices from supermarket websites to help with the inflation data. They probably will eventually find a way to get a reliable claimant-count figure from Universal Credit and maybe there is some hi-tech system we haven't yet thought of that will give more accurate, monthly figures for unemployment without having to spend millions of pounds extra on collecting them. As I mentioned when talking about surveys, the most accurate data would come from asking everybody in the country, but that takes ages and costs a fortune, so we need to accept that getting affordable figures reasonably quickly means they will not be absolutely precise.

Less helpful measures

Although the confidence intervals make the UK unemployment figures more difficult to report, they are at least clear and reasonably easy to understand, unlike an alternative measure: coefficients of variation or CVs for short. I hope you never have to deal with one of these, but just in case you do, a CV is an indicator of the quality of a figure – the smaller the CV the higher the quality. They are presented as percentages, and the true value is likely to lie within plus or minus twice the CV. Why would anyone create a measure like this? As an example, if you had a figure of 200 with a CV of 5 per cent, it would mean the actual figure is likely to be between 180 and 220.

I first came across these when I was head of statistics at BBC News and was asked to look at some figures that were the basis of a report due to go out on the *Six O'clock News* that evening. We had been sent figures by Labour suggesting that there had been big falls in average weekly earnings for women and the figures were broken down by constituency. They were based on a very big survey – the ONS's Annual Survey of Hours and Earnings (Ashe), which takes records on 1 per cent of employee jobs from PAYE records. But even the biggest surveys struggle when they are broken down into the 650 constituencies across the UK. Looking at an extreme case, the figures we were sent suggested that, in Putney, average weekly earnings for women had fallen from £460.60 a week in 2010 to £366.10 a week in 2013. That would be a staggering fall – about 20 per cent, which would be very unusual. It failed the reasonably likely to be true test, so I dug a bit further. Once you applied the CV to it, it turned out that we could actually say with 95 per cent confidence that the figure had fallen from between £343.20 and £578.00 in 2010 to between £222.60 and £509.60 in 2013. There is considerable overlap in those figures, so you can't say with confidence that wages had fallen at all. And as big as the suggested fall was, it was dwarfed by the range of the figures. That was the case for many of the numbers, especially the ones with the biggest falls that were being used as examples. We had to drop the story, which was a great shame because a lot of work had already gone into preparing it for the programme.

This story highlights another problem with wages figures, which is that they often don't tell you what you think they are telling you. In this case, we were being asked to use average weekly earnings to assess what happened to pay for women. It is a particularly unhelpful figure because women are considerably more likely to work part-time than men. If a new business had set up in the area and employed lots of women part-time but on generous hourly wages, the average weekly earnings figure would fall because the average number of hours people were working would probably fall. But that would not suggest women were being paid less. Average hourly earnings would have been more helpful or indeed a breakdown into full-time and part-time workers.

It's worth bearing in mind that we're only talking about the confidence intervals here because we're looking at some of the world's biggest and most robust surveys. Most surveys that you see, especially some of those we looked at in Chapter 1, would have margins of error off the scale.

But even when you're looking at the highest-profile statistics, it's worth considering whether the methodology is robust enough to say that something has risen or fallen in a particular month.

Making sure you see both absolute risk and percentage changes

I talked in Chapter 4 about the dangers of lonely percentages, which is the problem of being given a percentage change without the absolute figures. This is particularly a problem when you're talking about percentage risk of getting a particular disease, when there are real dangers of scaring people unnecessarily. Newspapers seem to have daily updates of what causes or prevents cancer, and it is very hard to decide whether you should be changing your behaviour without the full context.

It may feel like a bit of a handbrake turn to go from uncertainty in surveys to risk in health stories, but they are very much part of the same area of statistics – how likely something is to be accurate and how likely something is to happen.

In March 2008 there was a headline in the *Daily Mail*: 'Why eating just one sausage a day raises your cancer risk by 20 per cent'. If you read on further it turns out that the finding is that eating 50g of processed meat a day, which is one sausage or three rashers of bacon, increases your risk of getting bowel cancer by one-fifth. And then there's a photo of a schoolboy eating a sausage, although the caption points out that the picture was posed by a model, presumably because it would just be too damn dangerous to let a real schoolboy eat a sausage.

Four years later in the *Daily Express* there was a spookily similar headline: 'Daily fry-up boosts cancer risk by 20 per cent'. It's much the same story except that this time it's talking about pancreatic cancer instead of bowel cancer and they have used a close-up photo of some bacon to illustrate the story without exposing any models to processed meat danger. These headlines are scary – cancer is clearly a bad thing and 20 per cent is a big increase in the risk.

Communicating risk is not something that is done very well on the whole and the man trying to help us all to do it better is the aforementioned Professor Sir David Spiegelhalter, who runs a statistics lab in Cambridge that investigates better ways to help people understand about risk. He recommends looking at how much the level of risk has changed (the relative risk) in the context of the absolute risk. So we know that the change in the relative risk is 20 per cent, which is the difference in risk between people who don't have the extra sausage a day and those who do. But we also need to know the absolute risk: the actual numbers of people who could be affected by eating the sausage. With pancreatic cancer, 5 people out of every 400 will develop the disease in their lifetimes if they do not have that sausage a day. If they do eat a sausage or three rashers of bacon every day of their lives then that increases to 6 people. An increase from 5 people to 6 people is indeed a rise of 20 per cent, but the actual number of people sounds less alarming than the original headline. It may well be that when you understand the absolute risk you feel it is time to abandon daily bacon sandwiches, or you may feel the extra risk is worth it – we're somewhat outside my area of expertise here (I'm Jewish). The point is that you need both figures to be able to take that decision. If almost nobody is affected by a particular condition and that increases by 50 per cent then you are still left with almost nobody being affected.

Many of the other factors to look out for if you are scared by a health story in the news have been discussed in other chapters, but here are a few of the most relevant ones. I talked in Chapter 7 about questioning whether one thing is really causing another. You should also beware of reading too much into what a single study says. If it's an important piece of work then other researchers will try to investigate similar things and you will get a better idea of the quality of the work. Later, you will see research that pull together the results of many studies into similar areas, which will give a much more reliable picture.

I also talked in that chapter about randomised controlled trials, which require there to have been some sort of control over the experiment, generally allocating people to particular treatments at random. If the researchers have just observed what has happened or indeed looked back at what has happened without being involved, it is very difficult to demonstrate that one thing has caused another. You should be suspicious if researchers have not clearly decided in advance what they are looking for and may just be fishing in the statistics for something to investigate. And remember when you're talking about correlation to ask what else might be going on.

When you're looking at clinical trials in health stories, the number of people who should be involved is not necessarily the same as it would be in a survey. Clearly, having more people involved is better, because it makes coincidental outcomes less likely. But if you're studying a rare disease and there is a huge and obvious effect from treatment then it may be that just a handful of subjects will provide enough information from which to draw conclusions (although many more may be needed to get regulatory clearance). And it is important to check whether the headline is actually justified by the conclusions in the research, which I will discuss in more detail in the next chapter.

Another phenomenon that is important in this area is regression to the mean, which sounds complicated but is relatively straightforward. It's the idea that in situations in which there are many factors involved, an extreme result is likely to be followed by a more normal one. If your football team wins 9–0 this Saturday in the Premier League, you would expect a more normal one- or two-goal margin the following week.

The phenomenon is easily explained by a game for which I am indebted to former chair of the UK

Statistics Authority and *More or Less* presenter Sir Andrew Dilnot. He explained it in the context of a decision about whether to put speed cameras on a particular road. Everyone in the room is given two dice to roll, which represents the number of accidents on a particular road in year one. The people with the highest scores, generally 11 or 12, are declared to be accident blackspots. As a result, it is decided that speed cameras must be pointed towards them as they roll the dice a second time to find out how many accidents happen on their roads in year two. Of course, the number of accidents almost always falls, which vindicates the decision to use speed cameras. The point about regression to the mean is that in situations that may be affected by chance, such as the number of accidents there are on a road, an extreme year is likely to be followed by a less extreme one. In other words, unusually high or low numbers may return to normal just by chance.

The game may seem frivolous, but it is relevant to many decisions taken by governments. When they consider whether to take action to deal with a problem, it is likely that they will try to deal with the extreme cases. But it's possible that it's only a coincidence that these were the extreme cases, in which case you would expect a more normal outcome the following year.

This also works in health cases in which you are likely to treat the most seriously ill sufferers from a particular disease, some of whom may become closer to the average level of illness by chance even if the treatment you are trialling does not really work.

Regression to the mean was first discussed by Sir Francis Galton in 1877. One of the examples he used was the height of parents compared with their children. Tall parents tended to have children who grew up to be shorter than them, while short parents, on average, had children who grew up to be taller than them. This is because children will, on average, be average height, so you should not be surprised by the result of comparing their height with the height of very tall or very short parents.

Beware of extremes when judging measures taken by authorities. It is statistically difficult to exclude regression to the mean from a study, but if the researchers make it clear they are aware of the problem it is a good sign.

Deciding whether the odds of more than one thing happening are being considered correctly

An important question to ask when looking at the chances of more than one thing happening is whether the odds are linked or independent. If you flip a coin, the chances of getting a head are one in two. The chances of getting two heads in a row are one in four, three heads is one in eight, and so on. That is because the odds are independent so you just multiply them each time. The chances of getting tails are not related to whether you threw heads last time. But odds are not always independent – sometimes they are linked.

I was in the *More or Less* studio on the day of an item about double-yolked eggs. The British Egg Information Service tells us that the odds of finding an egg with two yolks is one in 1000. A BBC colleague had cracked four eggs in a row and found them all to have double yolks. If finding double-yolked eggs were like flipping a coin, then the odds of getting two double-yolked eggs in a row would be one in a million (a thousand times a thousand), three would be one in a billion and four would be one in a trillion, which is pretty unlikely. But that ignores the possibility that double-yolked eggs come in clusters, so that finding one in a box would make it more likely to find another. And that is indeed the case, because double-yolked eggs are more likely to come from hens of a particular age (20 to 28 weeks old), flocks of birds tend to all be of about the same age and the eggs in a box are all likely to come from the same flock. Also, double-yolked eggs will be unusually big for hens of that age, which means if you have bought a box of large eggs from a young flock, they are even more likely to have double yolks. The excitement in the studio was overwhelming as my colleague cracked the last two eggs from her carton and they both turned out to have double yolks, putting the odds against the streak at either a quintillion to one, or very considerably less, depending on whether you take into account linked odds.

This may not seem terribly important when it comes to egg yolks, but the failure to understand that some odds are linked was hugely important when it came to the financial crisis. US housing lenders had lent money to people who had a high risk of defaulting on their loans. Lots of those loans were then packaged together for investors to buy, but they were classified by ratings agencies as being low risk. How was that possible? Because the investments would pay out unless all of the mortgages in their product defaulted. Even if the risk of one borrower defaulting is relatively high, the risk of all of them defaulting is relatively low if you believe they are independent. What that ignored was the fact that the reason why one borrower might default could be linked to the reason why another defaulted, for example, a downturn in the economy or a huge housing bubble bursting.

The assumption of independent odds was also a mistake made in the tragic case of Sally Clark, who in November 1999 was convicted of the murders of her two babies and served three years in prison before being cleared. She never recovered and died in 2007 at the age of just 42. Her defence said that both babies had died of natural causes, probably sudden infant death syndrome (SIDS). One of the prosecution witnesses said that the odds of one baby dying of SIDS in Mrs Clark's household was one in 8543, and hence the chances of two babies dying of the same thing would be one in 8543 squared, which is a chance of about one in 73 million. The first mistake was to assume that the two deaths were independent. The causes of SIDS are not fully understood and you certainly cannot rule out the possibility of hereditary factors.

The Sally Clark case also raised what's called the prosecutor's fallacy, which is that because the odds against the innocent explanation are so high, the accused must be guilty. This ignores a number of problems, in particular that a mother murdering two of her babies is also extraordinarily rare. The Royal Statistical Society protested against the original judgment, pointing out that the court should not have concentrated on the odds against the innocent explanation, but weighed up the odds of the two competing explanations. Both explanations were very unlikely, but one of them had happened.

If we accept the one in 73 million figure, that's a figure trying to predict in advance whether two babies in one household are going to die of SIDS in their first year. But families were not being picked at random – Sally Clark was involved in this process because her two babies had died. So what was needed was not the odds that a random household picked from the population would suffer this tragedy, it was the odds that in a household in which this had happened, the mother would be responsible for both deaths. Given the lack of forensic evidence in the case, that certainly wouldn't have been high enough to prove her guilt beyond reasonable doubt.

Similarly, you sometimes hear claims that given the results of certain forensic tests, there is a 10 million to one chance that a particular defendant didn't commit the crime of which they are accused. Those sound like convincing odds, but only if there is some other reason to believe they did it. If they have just been picked at random, then at 10 million to one there could be six or seven other people in the UK population of 65 million who would get the same result in the tests. Of course, if the police find someone running away from the crime scene and forensic tests support the claim that they were the ones who committed the crime then it sounds like a strong case. But if the forensics lead police to arrest someone not obviously connected to the victim, that suspect could reasonably say that there is a good chance he or she didn't do it, and the judge has to be very careful in directing the jury in such cases.

So, at the end of a wide-ranging look at issues around uncertainty, risk and health stories, you now have the tools to avoid being taken in by insignificant stories. Look at whether the confidence intervals in a big survey mean that the changes in unemployment or migration that you're reading about are actually significant or are within the margin of error of the method being used to collect them. When you read scare stories about particular things causing a percentage increase in the chances of getting a particular disease, also find out what your risk of getting that disease without consuming those things would be. And when you hear about the authorities taking steps to deal with extremes, bear in mind that what goes up often comes down on its own – regression towards the mean suggests that situations will become average if they can, even without intervention. Finally, beware of the astronomical odds you sometimes see against something happening more than once – it may be that the two things are linked in ways that cut the odds.

CHAPTER 10

Economic Models

Decide whether you believe in them

On 18 April 2016 I was called into an early lock-in at the Treasury. A lock-in is what you get when something complicated is being announced at a particular time and an organisation wants to be able to explain it to journalists in advance so it can be reported correctly as soon as it's released. On this particular Monday, the Treasury was bringing out its long-awaited assessment of the cost of Brexit to the economy. We were briefed by a panel of somewhat uncomfortable-looking Treasury economists on what we could glean from the 200-page analysis in half an hour or so. The headline figure was that leaving the European Union and opting instead for a negotiated trade agreement would eventually knock £4300 per year off the UK's GDP per household. The whole lock-in was a bit of a mess because Chancellor George Osborne started his event in Bristol unveiling the analysis before the lock-in was finished, so most of the senior journalists who should have been watching it were instead locked in a windowless room in Whitehall.

The point of this chapter is not to discuss the Treasury's ability to coordinate a lock-in, but to talk about the sort of modelling that was used to come up with the £4300 a year figure.

Economies are very complicated. Economic modelling is a way of using past experience and theory to try to predict what will happen to parts of an economy in the future if particular things happen.

Getting to grips with how to dismantle an economic model is beyond the scope of this book, but there are plenty of questions you can ask without an economics degree. This chapter will cover three key questions you can ask:

- Is the conclusion justified by the model?
- Has a selection bias affected the outcome of the model?
- Are the assumptions that have gone into the model reasonable?

Is the conclusion justified by the model?

When we finally got out of the Treasury lock-in and reached a television, it turned out that Mr Osborne was in Bristol in front of a poster saying '£4,300 a year – Cost to UK families if Britain leaves the EU'. It was an interesting choice of headline figure. The conclusion of the research was that GDP (that's the value of everything produced by the UK economy) would be £120 billion a year lower in 15 years if the UK left the EU than it would have been had it stayed in. This is clearly a meaninglessly big pile of money, so the Treasury decided to divide it by something to make it more manageable. It could have done what the Leave campaign did with the EU contribution figure and divide by 52 to get a figure of about £2.3 billion a week, which would have the advantage of being easily comparable to the £350 million a week on the side of the Leave bus. Instead, the Treasury decided to divide by the number of households in the UK to get its £4300 a year. But here's the problem – a fall in GDP per household is not the same as a cost to each family. There is a relationship between the two, but it's not necessarily pound for pound because they are not the same thing. So a £4300 fall in GDP per household would cut household incomes, but not by as much as £4300.

I have since been told that the sort of model that the Treasury was running could have been used to come up with a household income figure if that was what was wanted. And indeed, the impact on household income could have been greater than £4300 per household because the falling pound would make imported goods more expensive, which hits the purchasing power of families but does not necessarily reduce GDP. But that wasn't what the Treasury did.

I published something on the BBC News website explaining the slip between GDP per household in the report and household income on the poster, which spoiled the day of Craig Oliver, Prime Minister David Cameron's director of communications, who wrote about it in his book about the campaign. I even had a phone call from a Treasury press officer trying to persuade me that GDP per household is the same thing as household income because all money eventually comes from households. But this is clearly not the case. GDP is currently about £2 trillion a year. If you divide that by the 27 million households in the UK you get about £74,000. But average household income is considerably less than that, so there is no question that these are different figures.

This wasn't a problem with the Treasury's analysis, it was a problem with what it had put on the poster. The first thing to check when you're looking at an advert, a poster or a newspaper report based on economic modelling is whether the top-line is justified by what the report actually says.

A good starting point when considering whether the claims are justified by the research is that the stronger the claim, the more likely that it's been exaggerated somewhere between the researchers and the press office. If you see a really big claim it should be sending you straight to the original research to see if there has been a slip-up between the researchers and the PR department. This happens disturbingly often because the job of promoting an institution's work may be at odds with the instincts of the people who have carried out the original research.

There was an example of this involving the *Journal of the Royal Society of Medicine* in 2017. It put out a press release saying: 'New analysis links 30,000 excess deaths in 2015 to cuts in health and social care.' Excess deaths is an estimate of the number of people who died in addition to the number who would have died anyway. If you're looking at excess deaths in winter you take a baseline of the number of people who would have died in a normal summer. Excess deaths may be due to cold weather or a particularly bad flu outbreak. The ONS brings out figures for excess winter deaths every year and there was indeed an unusually large number of excess deaths in winter 2014–15. The paper in the journal looked at some of the explanations given for the number of excess deaths and decided that they did not conclusively explain the rise. The researchers concluded in the paper that they were 'not able to reach a firm conclusion about what has happened' but added that 'the possibility that the cuts to health and social care are implicated in almost 30,000 excess deaths is one that needs further exploration'. So, the researchers had looked at why many more people had died that winter than usual, found that the reasons currently given for them were not convincing and suggested that one area worth looking into in future would be the cuts to health and social care. That translated into a press release saying that the analysis had linked the deaths with the cuts, which in February 2017 then spawned a banner headline on the front page of the *Daily Mirror*: 'Tory cuts killed 30,000'. You do not need any special qualifications in medical research to see that the conclusions of the paper did not justify the headline. Academic papers are handily structured to make them easy to check, with sections marked 'abstract' and 'conclusions' to help you understand the gist of what the research found without having to read the whole thing.

Has a selection bias affected the outcome of the model?

Selection bias sounds scary but is a fairly simple concept that you can use to impress your friends once you understand it. When you start looking out for it you will see it all over the place.

Selection bias is important in economic modelling and many other statistical areas. It is what happens when you're trying to collect data about an individual or group, but the way you are choosing is not properly random. If your economic model is based on choosing people in a way that could be skewing the results then it doesn't matter how well the rest of the model is designed, its results will be of little value. For example, I was asked to look at responses from a huge number of people who used a particular pension scheme and were answering questions about what they were going to do following a change in the rules for pension savings. The problem was that the people saving into the fund had far more pension savings than the average person in the country, so there was an immediate selection bias – all the people responding were unusually wealthy, so their responses could not tell you anything about the population as a whole. Selection bias is essentially the same as the question, raised in Chapter 1, about whether the right people are being surveyed.

Another example of selection bias was revealed in 2018 when a government minister came on Radio 4 and claimed that because people who are given community sentences are less likely to reoffend than those who are given short prison terms, community sentences should be used more. But clearly that is not a fair comparison, because judges will be more likely to give community sentences to people who they think are safer to be left in the community, which means they are considered less likely to reoffend.

Another good example was the research into weekend deaths in the NHS in 2015, to which the then Health Secretary Jeremy Hunt regularly referred when trying to get more doctors into hospitals at weekends. The study found that people admitted to hospital at the weekend were more likely to die, but warned against treating those figures as referring to avoidable deaths. It pointed out that routine operations are not carried out at weekends, so people will only be admitted to hospital then if there has been some sort of emergency. That means that those admitted at the weekend are likely to be sicker than those admitted on weekdays, which may explain the disparity.

Avoiding the effects of selection bias is important when you're trying to build an economic model based on the impact of something happening. For example, imagine you're trying to find out whether trade deals are good for an economy. Which trade deals would you look at to make that decision? If you were doing this in 20 years' time you could look at the deals done by the UK following Brexit, but the selection bias there would be that the first deals done would not be random – it would make sense to do the first deals with the countries with which the UK does the highest volume of trade. Alternatively, the first deals might be the ones that are the easiest to do or ones that maintain arrangements that are currently in place as part of the UK's membership of the European Union. In any case, it would be very difficult to identify a 'normal' deal. And the impact depends on many factors such as how far apart the countries are and how wealthy they are. All this makes building the model very difficult.

Imagine you were trying to find out what effect migrants have on jobs or wages. A way you could find out would be to look at an area that lots of migrants had moved to and see what had happened to jobs and wages there. But people do not choose places to move to at random – they will tend to go to places where there are lots of jobs and decent wages, so there is an automatic selection bias.

I talked about randomised controlled trials in Chapter 7 and they are a factor again here. If you wanted to find out the impact of migration on jobs and wages you would have to randomly choose to send one group of people to an area where there were lots of jobs available and one group to a place where there were not and see what happened. People tend not to volunteer for such research, especially if it means they might get sent to try to find employment somewhere that jobs are scarce.

An alternative is to use what are called natural experiments, for example, seeing what happens when political upheaval or conflict leads many people to move in numbers to a country for reasons other than a favourable job market. The movement of Germans from east to west following the fall of the Berlin Wall is an example of this, as is the movement of French expatriates from Algeria in the 1960s. But even these natural experiments throw up other factors, such as companies moving to the places where the migrants are arriving to take advantage of the supply of workers.

As with the question of correlation and causation in Chapter 7, the other question you need to ask yourself when considering an economic model is what else is going on here? So, for example, if you are looking at research into whether babies who are breastfed for longer tend to go on to have better outcomes in education or health, bear in mind that, at least in developed countries, mothers from higher-income backgrounds tend to breastfeed for longer and so their children already have a head start.

Selection bias crops up a lot when choosing samples for surveys, but you can see them in responses to almost anything. You probably received many emails in 2018 about the General Data Protection Regulation (GDPR) asking if organisations you don't remember ever having contacted could keep you on their databases. If you're like me, you probably replied to one or two of them straight away that you were particularly interested in and ignored the rest. If you were running the GDPR policy at an organisation, you might look at the responses you have received on the first day after sending out your email and find that 5 per cent of people had responded and three-quarters of them wanted to continue to be contacted by you. That's a perfectly reasonable statistic unless you try to do something else with it. Can you conclude that three-quarters of your current contacts want to continue to hear from you? Of course not – the responses on the first day are the people who are unrepresentatively interested in what you do. It may be that you don't have any more responses at all beyond the first day. From what I'm told by friends involved in making their organisations GDPR compliant, that that would not be unusual.

Darrell Huff in *How to Lie with Statistics* gives the example for selection bias of what would happen if you sent out a questionnaire to thousands of people asking them if they enjoyed filling out questionnaires. How many people do you think would bother returning the questionnaire to tell you that they didn't?

When you're looking at a model consider what sort of effect selection bias may have had on it. Is the effect that they are trying to model a difficult thing to measure? In particular, does it involve people making decisions? People making decisions almost always introduces selection bias, whether it's a question of if they should move to another country to work, whether they should continue to breastfeed their baby or whether they can be bothered to reply to an email.

Are the assumptions that have gone into the model reasonable?

We have a saying at Reality Check, the BBC's fact-checking team, that you can't fact-check the future.

People make predictions about things all the time – they might be right and they might be wrong. They might be right by coincidence, in ways that are in no way related to the steps the researchers took to make their predictions. Not only do we not currently know whether the predictions are right – we will almost certainly never know. Somebody making warnings about some terrible thing that will happen in the future may claim that, as a result of his warnings, action had been taken to prevent the terrible thing from happening.

The £4300 a year in the Treasury report was not a statistic – it was a forecast based on economic modelling. There are ways to check whether you think a piece of economic modelling is basically sensible, but before you do you need to take a view on whether you want to be influenced by economic modelling at all. Organisations that use it to make predictions do so because they have to in order to inform their decisions or plan their spending. They come up with their best estimates and take decisions accordingly. It's not just governments that do economic modelling. When you see an advert encouraging you to support a charity on the basis that food prices are going to double, for example, that will be a conclusion based on economic modelling.

If you are going to be influenced by it, let's start by saying that it's a tricky thing to do. The modelling carried out during the EU referendum campaign to establish what impact leaving the EU would have was particularly tricky because there were so many unknowns. We didn't know what sort of trade deals the UK would manage to do after leaving, how long they would take to negotiate, how much of the UK's contribution to the EU Budget would be saved, what that saved money would be spent on, whether the regulations that the UK government devised to replace the EU ones would be better than the EU ones and what effect all that would have on the economy. We also didn't know whether Brexit would create some sort of feel-good factor in the UK economy, or the opposite.

But what made it particularly hard was that models are built based on what particular events have done to economies in the past. We know that joining free-trade areas tends to increase trade and economic growth, but we don't know what effect leaving free-trade areas has because it hardly ever happens. Economists trying to model this looked as far back as the break-up of great empires in history to work out this effect. In the end, most of the models around the referendum were based on the idea that if joining free-trade areas was good for trade and growth then leaving them was likely to be the reverse.

The Treasury used what's called a gravity model, which considers trade between countries and models what happens to it based on factors such as how far apart they are, whether they use the same language and how rich they are. It then looks over time at what happens if some of these variables change. But you have to start out with various assumptions when you build the model, such as what happens if currencies weaken or how much signing a free-trade agreement with one country reduces the amount of trade with other countries with which you do not have an agreement. And these models are enormously sensitive to the assumptions on which they are built and the scenarios that they are supposed to be judging.

The £4300 per household figure came from the Treasury's long-term analysis, but the department also conducted short-term analysis. It suggested that a vote to leave would create an immediate and profound economic shock, push the country into recession and lead to a sharp increase in unemployment. This clearly has not happened. There are a few reasons why the predictions were wrong and the first is that David Cameron had said that following a Brexit vote he would immediately go to Brussels and trigger Article 50, which starts the process of leaving the EU. In the event this didn't happen until the end of March 2017, which was almost a year after the vote. The second problem was that it assumed that a vote to leave the EU would cause shocked consumers to reduce their spending, but missed the idea that if consumers had voted for it they probably wouldn't think it was a bad thing. Thirdly, it assumed that the government and Bank of England would not take any action to support the economy, when in fact the Bank of England cut interest rates and pumped extra money into the economy. Would you have accepted these assumptions? It was fair enough to believe that the Prime Minister would trigger Article 50 straight away. Many commentators at the time pointed out the assumption that the authorities would not take steps to support the economy. I did not see anybody point out the flaw in the assumption about consumer spending falling.

The Treasury was right in its forecast that the pound would fall sharply, but wrong in many of its other predictions of the immediate outcome. This does make me wonder whether it was a good idea for the Treasury to be involved in this sort of forecasting at all. When I asked a statistician in the civil service if he was worried that departments putting out these sorts of forecasts reduced the public's confidence in other things they publish such as official statistics, he told me that the forecasts come from the economists while the statistics come from statisticians. I'm not convinced that people recognise this distinction. The establishment of the Office for Budget Responsibility to make official forecasts for the government and keep politics out of economic forecasting was an excellent idea. Perhaps it would be a good idea if the government took advantage of this to avoid getting involved in forecasting at all. The UK's statistics regulators are trying to widen their code of conduct to include things such as economic modelling that are not actually statistics in order to prevent damage to the reputation of government departments from forecasting and other numerical releases.

This distinction between what statisticians do and what economists do is quite important although not precise – there are economists involved in the creation of official statistics just as there are statisticians who help with models and forecasting. Even so, I find it strange that almost all news organisations have economics correspondents and, as far as I know, only two have a statistics editor.

One is at the *Financial Times*. I met its head of statistics when the BBC first decided it was prepared to fund a statistics role for a trial period. I stole his title, but I did ask him first. The BBC now has a permanent head of statistics, in addition to its economics editor and correspondent, which I think has made a considerable difference to its reporting of numbers.

Should you be ignoring what economic models say? George Box, one of the greatest statisticians of the twentieth century, said: 'All models are wrong, but some are useful.' The precise numbers may be of little value, whether they are the £4300 a year figure or the Treasury's earlier declaration that every person in Scotland was £1400 a year better off inside the UK, but the direction the models predict and some of the assumptions they make are quite interesting.

My friends at the fact-checking website Full Fact

offer the analogy that you would probably listen if your doctor told you to stop eating junk food because it would make you fat, even if she couldn't tell you exactly how many kilograms you would weigh this time next year. If you buy into the assumptions then you are likely to accept the direction of travel.

As with surveys, a good first step when looking at an economic model is to look at who has conducted the research and who is paying for it. If it has been commissioned by a campaigning group you are justified in being a bit suspicious of its findings, but remember that an independent group can be just as wrong as a biased one.

Next, take a look at some of the assumptions on which the models are based. Reports in this area should be transparent about their assumptions and how the model works – if they are not you shouldn't believe them. If they are then you can take a view on whether you think they are too optimistic or too pessimistic. Also, have a look at what the report says about the level of uncertainty – the Treasury's reports were pretty good at highlighting the uncertainty in their models. Anyone claiming to know exactly how things will pan out in the future is lying to you. And remember that if you are not in a position in which you need to use economic modelling to take decisions, you could choose to ignore it altogether.

There is a middle ground here. You may take a view that the numbers you end up with as a result of an economic model are not particularly helpful, but perhaps the assumptions it uses will help you get back to first principles. The first principles for the Treasury model would be that they believe trade is a good thing because it allows the country to specialise in things that it is good at doing while importing products from other countries that are better at producing them. It could then have argued that EU membership makes trade easier with the UK's nearest neighbours and some of the world's richest economies. I wonder if going back to first principles without the numbers would have worked – people working in senior positions in both campaigns tell me they were under great pressure from the media to produce numbers and not just arguments, so perhaps not.

As long as this pressure exists it is important that you recognise what you can do to challenge the numbers being thrown at you from models. Check whether the claims are justified by the research and look at the assumptions being made. Then take a view on whether the numbers are any use to you and if you wouldn't be better going back to first principles to understand what is going on. Also, think about whether the question being asked in the modelling has introduced a selection bias – if anybody is having to take a decision then it probably has, and you need to look at how the economists have allowed for that. And if all else fails, you may be able to just ignore the models and still get on with your life.

CONCLUSION

But I really needed that figure

I'm sometimes told by colleagues that if they followed my advice then we would have nothing to report. Part of me feels that if everything available to report would have been ruled out by the ten chapters you have just read then cancelling the news and showing a selection of classic cartoons instead would be no bad thing. On 18 April 1930 in the 8.45 p.m. BBC news bulletin, the announcer declared 'There is no news' and then piano music was played for the rest of the 15-minute segment. It's hard to imagine that happening today.

But, of course, a commitment to statistical robustness does not mean abandoning all stories, it means reporting them correctly, and because you're looking back to the original research rather than just repeating what the press release says, you get a different story and generally a better one than all your competitors are running.

This is true for all uses of numbers, not just news

coverage. Most statisticians hate the quote that Mark Twain, probably incorrectly, attributed to Benjamin Disraeli, that there are three kinds of lies: lies, damned lies and statistics. Part of the reason for this is that it's hard to imagine how statistics can lie – they may be inaccurate, but I think that in order to lie you need words. Often it is not the numbers that are important: what really matters is the words around them.

This is particularly the case if you find yourself needing to use the best-available statistics, even when they are not as good as you might hope. The BBC's *Victoria Derbyshire* programme approached me to help with some research its team wanted to use about the experiences of male sex workers. Researchers had managed to get the responses of about 120 male sex workers, but this is not an easy group to survey and we believed it was the biggest sample anybody had managed to get. You know from Chapter 1 on surveys that 120 is not nearly enough to make general statements about all male sex workers, but if you choose your words carefully this is still research that can be used.

I am a great supporter of using the best research available, but only if you tell your audience what you are doing. Remember the rule of thumb that if you can explain out loud in detail where the figures have come from and what they cover without feeling foolish then they are likely to be OK. In this case, you can say that a survey of 120 sex workers found that there was considerable under-reporting of crimes against them, and hope that anyone reading it would understand that this was not an easy group to survey.

Similarly, while you now understand that costings are not an exact science, there are some areas where the best estimate is useful. If you're a politician trying to decide whether to give the go-ahead for a big infrastructure project or you are taking an investment decision at work, you are probably going to need some figures to help you decide. Evidence-based policymaking is a good thing. But just consider whether the project would be a very bad idea if the costings you have been given turn out to be drastically wrong. Give yourself a healthy margin of error around estimates of costs and benefits to come.

In that same situation, you are likely to be presented with economic models to help you take decisions. Look under the bonnet of the models and see what the estimates are based on. There's probably some useful stuff there to help with your evidence-based policymaking, even if you are dubious about the overall conclusions. Consider what the risks are if you follow the advice based on a particular model and it turns out be wrong.

When you come to tell people what decisions you have taken, always choose your language carefully to get across the uncertainty in the figures you are using. Remember that almost no government statistics are based on counting things, they are mostly estimates based on taking a sample – make sure your words reflect that. It's only when you really understand what's wrong with certain figures that you can be in a position to use the best-available figures properly. A few years ago a colleague from the BBC's Persian service asked for some advice on reporting the Iranian budget on the basis that he did not believe all of the figures in it. The answer was not to put questionable numbers in the headline or high up in the story and, most importantly, never put numbers that you are not confident about in a chart. Readers tend to believe anything in a chart passionately, however much you tell them in the text that the figures are a bit suspect. This has been my experience, and there was also some research done at Cornell University in the USA that found people were more likely to believe something if it was in a graph.

Look out for the alarm-bell words and phrases, check whether the things that correlate are really causing each other, and look out for the ways people might be trying to mislead you with averages and percentages.

Fundamentally, we do not have answers to many of these problems, but if you are aware of them then you are well ahead of the game. If you understand the flaws in some statistics then you will be able to pick the words that allow you to use the best-available data, or recognise when somebody else is failing to do so.

If you have gained one thing from reading this book, I hope it is the confidence to challenge numbers in the same way as you would challenge any other evidence. If you can do that then you will genuinely be in a position to decide whether a number, a claim or a news story is reasonably likely to be true.

Acknowledgements

I owe so much to my late father, Professor Bryan Reuben, who taught me to ask whether things are reasonably likely to be true. We had been planning to write a book together about why costing is bogus when he died. I am also indebted to my mother, who has taught me so much, and in whose house I wrote this book. My wife Susan has given love, support and suggestions throughout the process, not to mention actually reading the book and making useful comments. My children Isaac, Emily and Boaz have taught me that I always need to concentrate when answering questions.

My agent Ben Clark from LAW has been extremely enthusiastic about the project from the start and coaxed me through the process with great skill and understanding. As soon as I met my editor Claire Chesser from Constable I knew that I wanted to work with her – she's great. Howard Watson copy-edited the book and is the first person who has ever suggested that I use too many adjectives. I was delighted.

I have been promoting statistical robustness at the BBC for at least a decade with the help of many very special people, especially Jonathan Baker, who managed to find funding for me to spend 18 months as the corporation's first head of statistics. That mantle has now been taken on by Robert Cuffe who has proper statistical qualifications and is learning to be a journalist with alarming speed.

Building the Reality Check brand has been enormous fun and would not have happened without Jonathan Paterson, Alexis Condon, Tamara Kovacevic, Rachel Schraer, Peter Barnes, Tom Edgington, Juliette Dwyer, Liz Corbin, Rupert Carey, Chris Morris and a whole range of scarily knowledgeable researchers. Thanks to my friends at the Office for National Statistics, especially those in the press office who have made me the most spoiled journalist in the country. Also Glen Watson, the now retired director general of the ONS, who was a huge support in my early work as head of statistics and lent me first Jamie Jenkins and then Steph Howarth to help. I'm also grateful to everyone at the Royal Statistical Society, which has been very supportive throughout.

Thanks to Marc Webber for the joke about making up the numbers, to Richard Posner for pointing me towards the reporting of crime statistics in Nottingham and to Jen Clarke for finding all the double yolks. Daniel Vulkan, Sarah Lowther and Nick Blain also pointed me towards examples I could use.

Several people have generously agreed to read drafts of all or part of this book and made helpful suggestions. The first to read it were my brother David, who is the most numerically pedantic person I know, and Adi Bloom, who is the most grammatically pedantic. Corinne and Ben Sheriff, David Cowling, David Sumpter, Robert Cuffe and Malcolm Balen also gave me the benefit of their expertise and made useful suggestions.

Any errors are of course my fault – I look forward to an eagle-eyed reader finding one.

Malignant Sadness
The Anatomy of Depression

LEWIS WOLPERT

faber and faber

First published in 1999
by Faber and Faber Limited
3 Queen Square London WC1N 3AU
This edition first published in 2006

Photoset by RefineCatch Limited, Bungay, Suffolk
Printed in England by Mackays of Chatham plc, Chatham, Kent

A CIP record for this book
is available from the British Library

ISBN 978-0-571-23078-5
ISBN 0-571-23078-4

2 4 6 8 10 9 7 5 3 1

Contents

Acknowledgements

Giving thanks is little recompense for all the help I have received. Maureen Maloney typed the manuscript and a first draft was read by Professors Martin Raff and Hugh Freeman, who made invaluable suggestions. Cynthia Kee listened to me read the whole manuscript to her and helped to improve it. My publishers, of course, played a key role: Julian Loose, my editor, and Robert Potts, my copy-editor. My agent Anne Engel was always encouraging, but it is too late to thank my wife Jill Neville for making me write the book in the first place. Finally, I am indebted to the authors of a large number of papers and books on the science of depression whose individual contributions I have not acknowledged.

L.W.

Introduction

It was the worst experience of my life. More terrible even than watching my wife die of cancer. I am ashamed to admit that my depression felt worse than her death but it is true. I was in a state that bears no resemblance to anything I had experienced before. It was not just feeling very low, depressed in the commonly used sense of the word. I was seriously ill. I was totally self-involved, negative and thought about suicide most of the time. I could not think properly, let alone work, and wanted to remain curled up in bed all day. I could not ride my bicycle or go out on my own. I had panic attacks if left alone. And there were numerous physical symptoms – my whole skin would seem to be on fire and I developed uncontrollable twitches. Every new physical sign caused extreme anxiety. I was terrified, for example, that I would be unable to urinate. Sleep was impossible without sleeping pills: these only worked for a few hours, and when I woke up I felt worse. The future was hopeless. I was convinced that I would never work again or recover. There was the strong fear that I might go mad.

I had never been seriously depressed before. On previous occasions the way I dealt with mild depressions – feeling low – was to go jogging. Enquiry among my fellow joggers confirmed my view that we do not exercise for health but to avoid mild depression. The widely held belief that exercise raises endorphin levels and so provides an uplift in mood turns out to be based on quite reasonable scientific evidence. I have to admit that I then rather sneeringly proclaimed that I believed in the Sock School of Psychiatry – just pull them up when feeling low. But that certainly does not work with serious depression. The origins and course of my own depression, and my recovery from it, will be described in later chapters.

My wife, Jill Neville, was embarrassed by my being depressed and told colleagues and friends instead that I was exhausted from a minor heart condition. She was worried that if the truth were known it would affect my career. When I recovered, I was most uneasy about the stigma associated with depression, and the shame felt by many sufferers; it seemed to me a serious illness of which one should not be ashamed. I therefore decided to make my depression public and wrote an article about it in the *Guardian* newspaper. This brought an astonishingly positive response. Patients, doctors and those who had had the experience of living with someone who is depressed found it helpful to have the subject discussed in so open a manner. Of everything I have written, both books and scientific articles, this article was most widely read and appreciated. When people complimented me on being so brave, I realised exactly how much stigma is still associated with depression. In fact it was quite easy for me to write about since I had a secure academic position and so nothing to lose.

After I had emerged from my depression I thanked the psychiatrist who had treated me for all her help. I then asked her if I was correct in thinking that psychiatrists really understood nothing about depression. She partly agreed. Of course they have great skills at diagnosis and treatment; for example, antidepressant drugs like Prozac can bring about remarkable recoveries. But it was at a mechanistic level that little seemed to be known. It was even far from clear to me what it meant to 'understand' a mental illness, in the same way that one now understands cancer. For example, we can understand cancer in terms of the changes in certain genes involved in the control of cell multiplication, and also in terms of the spread of the malignant cells. But even if low levels of serotonin, one of the chemicals in the brain linked to depression, were found to be in some way responsible for the illness, this alone would still be inadequate as an explanation. For how could changes in the concentration in the level of so simple a molecule bring about such profound changes in behaviour as are experienced in depression?

Although there are many 'self-help' books on the subject, I found very little reliable information about depression easily available, and decided to write this book to set down what is known. My purpose is fourfold: to help those who are living or working with a sufferer to understand the nature of depression, since depressives, whether parents, children or companions, are not easy to be with; to help depressives to understand themselves; to remove the stigma associated with depression; and, foremost, to try and understand the nature of this dreadful affliction in scientific terms. This last aim is something of a personal quest.

I know that I am entering into areas where I have no direct expertise, being neither a doctor nor a psychologist, but I do have two advantages. I am a research biologist whose interest is in the mechanisms by which embryos develop and the way that genes control cell behaviour and generate limbs and other organs, so I am familiar with basic biological processes and complex systems. As a scientist I also have some experience of assessing evidence. But more importantly, I have experienced depression, for anyone who treats or writes about depression and who has not themself been depressed is rather like a dentist who has had no experience of toothache.

Depression is very upsetting not only for the sufferer but for those who live with the victim. Depressives are victims in the sense that they have a frightening and disabling illness; an illness that affects as many as one in ten of the population and is twice as common in women than in men. Considering how widespread depression is, it is particularly unfortunate that it carries with it the additional burden of severe social stigma.

The effect of depression on health-care services is enormous. A recent report, *Global Burden of Disease*, published by the World

Health Organisation, states that depression was the fourth most important health problem in the developing world in 1990 (accounting for about 3 per cent of the total burden of illness) and predicts that it will be the number one health problem in the developing world in 2020 (accounting for about 6 per cent of the total burden). Over the same period the annual number of suicides will increase from 593,000 to 995,000 in the developing world. The report also estimates that less than 10 per cent of the 83 million episodes of depression in the developing world in 1990 received treatment and that the figure for treated episodes in developed countries may be only two to three times higher.

Depression has a confusing number of different meanings. In common usage it refers to lowness and anxiety, common feelings in everyday life. But it is depression as an illness with which this book is concerned, depression that so interferes with a person's life that it is disabling. William Styron's *Darkness Visible* is a marvellous description of depression, and at the very start he makes it clear that the 'pain of severe depression is quite unimaginable to those who have not suffered it, and it kills in many instances because it cannot be borne'. So the focus in this book is on major depression, or, as it is so often called, clinical depression; depression so severe that it can lead to the inability to work or even to suicide. The relationship between major depression and common everyday depression, just feeling low, is, however, an important one and will be explored: is major depression just an extreme form of common depression or is it qualitatively different?

My title is in two parts. One comes from Robert Burton's famous, monumental, fascinating, but not easily readable, *Anatomy of Melancholy* (1621–51) in which he recorded all aspects of the melancholic condition known at the time. Burton spent most of his life at Christ Church in Oxford where among other duties he taught theology. He had an interest in all branches of medicine and science. He chose to write about Melancholy as his life's work largely because of his own affliction by it, and he hoped that writing about it would alleviate his symptoms: 'I write of Melancholy, by being busy to avoid Melancholy.' As the choice of Anatomy for his title implies, he tried methodically to exhaust the topic and cite every known authority. Burton also cared about the style of his writing and would have been gratified had he known that Samuel Johnson, himself a depressive, turned to the *Anatomy* for consolation – it was the only book that ever took him out of bed two hours sooner than he wished to rise.

The number of papers published about depression is currently more than 3,000 every year, so I have had to be less ambitious than Burton. The amount of information is enormous, but I try to summarise in an accessible form what is currently known about depression. I start by looking at the experience of depression in the past and present. Then I look at the problems of diagnosis not only in the West but in other cultures. I try to unravel the factors that make people vulnerable to depression, such as their genes, distressing life events, early childhood experiences and even the weather. Manic depression, though not central to this book, has its own characteristics, and suicide has to be recognised as a tragic consequence of depression: I address these subjects in separate chapters. With this background, it becomes possible to discuss the psychological and biological theories that have been put forward to explain depression, including its evolutionary significance. The psychological explanations focus on the importance of loss and early experience, while the biological require understanding of emotion in terms of brain function and chemistry. Following this, there are discussions of the treatments for depression, such as medication and psychotherapy, with an analysis of what works and for whom. I also report on experiences in the treatment of depression in the East – China, Japan and India. Finally I look to the future, both at scientific advances and preventative approaches.

I have been particularly influenced by several books, including William Styron's *Darkness Visible*, an account of his own depression; Kay Redfield Jamison's *Touched with Fire*, which deals with manic depression and creativity as well as other topics related to depression; and *The Emotional Brain* by Joseph Le Doux. Several ideas have also been very influential on my approach, particularly John Bowlby's ideas about attachment and loss and Aaron Beck's ideas on the cognitive basis of depression and its relation to negative thinking. Arthur Kleinman, an anthropologist and a psychiatrist, has illuminated for me the nature of depression in other cultures.

The main title, *Malignant Sadness*, is meant to emphasise the very serious nature of a depressive illness and also to reflect my conviction that normal sadness is to depression what normal growth is to cancer. I hope this book will prove interesting and helpful both to those who suffer from depression and to those who live with them.

(1999)

I am continually amazed how widespread depression is. Because I have gone public about my depression, it is very rare for me to meet someone socially who during discussion does not reveal that they have had some contact with depression – a relative, a friend or even themselves. People who would not normally talk about their depression felt quite comfortable talking to me, for they knew I had had a similar terrible experience. Many shared my view that being seriously depressed was quite unlike anything one experienced outside of depression. Terms like 'black hole', 'a different world' and 'indescribable' are often used.

The response to the publication of this book was very gratifying. Most satisfying were direct letters and direct contacts. I was grateful for and moved by the many letters I received. Almost all were complimentary and thanked me for describing my own experience, making it public, and sometimes even helping someone deal with their own depression. Encouraging too were those who cared for someone who is depressed and who said that the book helped them understand this, to an outsider, mysterious and distancing illness. I also received calls for advice which I was not competent to offer. Always my response was to urge them to get professional advice as soon as possible.

There were some very good and positive reviews from reviewers for whom I had great respect. Not all reviews were positive, the main complaints being that I did not write enough about myself, and that I was hostile to psychoanalytic interpretations and too committed to biological, even scientific, interpretations.

The stigmatization of depression is very disconcerting. I am often congratulated on being brave, even courageous, in talking so openly about my depression. This is a clear implication of the associated stigma. In fact, for me, there is no 'bravery' at all. Public writing and lecturing are part of my everyday work, and I like to think I feel no shame whatsoever about having been depressed. But perhaps this is not quite true, for, as I shall explain, I much prefer a biological explanation for my condition to a psychological one.

For many others the stigma of depression is a serious problem, and there were those who told me that they could not confide in their brother or sister even though they had attempted suicide. Yet others were convinced that if their condition were known they would lose their job. Some were very successful, surprisingly, at keeping their depression hidden when in public. I remarked to one woman with whom I was lunching that she seemed, in spite of her condition, remarkably cheerful. She told me how she had learned to conceal it. It was chilling to hear her describe how cheerful she could be with her son at the same time that she was composing, in her mind, the suicide note she would leave him.

Repeatedly I was told stories of people who were seriously depressed but refusing to take an antidepressant. There seems to be a widely felt anxiety that antidepressants are addictive and that they will change or damage one's mind. A chemical solution arouses suspicion and hostility. Yet St John's Wort is more acceptable. Irrational though it may be, I am sure that if Prozac grew on trees people would be much more willing to take it.

I liked to believe that my depression had a purely biological basis and was induced by a drug I was taking for my heart. That this conviction had no psychological basis made me suspicious, but I stuck to it. Even though I publicly declared – was almost evangelical – that depression should carry no stigma, this conviction somehow made me feel more comfortable. A biological cause implied that I was not responsible. It was not unlike having a diagnosis of post-traumatic stress disorder. This is probably the only psychiatric diagnosis that carries no stigma – because the cause is included in the diagnosis, and the condition has an external and well-defined cause for which the individual cannot be held responsible. In a similar way, I was sure that my depression was caused by the drug.

How wrong I was. While I knew, and had written in the book, that only one in ten patients who have recovered from depression will not have a relapse, I believed that it would not apply to me. But nearly four years after I had recovered from my severe depression, I began to feel things were not quite right. The feelings were not specific. I had just finished writing this book and had done a three-part TV series for the BBC. Perhaps, I thought, it was coming down from the high of all that intense activity; a bit like post-natal blues. I went to South Africa for Easter and all was lovely. But on my return I found it difficult to work as hard as I am normally able to do. I had a quite unpleasant urinary infection in June and blamed it for my continual fatigue. That, I told myself, was what infections did. As a hypochondriac, I had earlier persuaded myself that I had diabetes or cancer, for I was starting to lose weight. A holiday in Crete was OK, though I lacked energy and slept a lot.

On returning I had many commitments. I went to Cracow in September and felt unwell and anxious. I began to have those low feelings, so very difficult to describe, that were unpleasant and frightening reminders of my earlier depression. I decided to take St John's Wort, and though it made me a bit nauseous I began to feel better. I was travelling a lot and was getting very tired. The feelings of an impending downward path returned, and I contacted the cognitive therapist who had helped me during my previous depression. She thought I was worrying excessively about my impending retirement. I visited her every two weeks, even on the morning of the big seventieth birthday party I had organized for myself. The following day I went to Holland for a lecture tour, and returned exhausted several days later. Other problems emerged – my prostate was enlarged, making me get up almost every hour during the night, and the prospect of an operation to deal with it seemed unavoidable. Bad news even for a non-hypochondriac.

Then I started waking up early in the morning and being unable to go back to sleep. This was for me a sure sign that my depression was closing in on me. I also found it very difficult to get up in the morning, and I could hardly keep awake in the afternoons. I was becoming increasingly anxious. I stopped taking the St John's Wort and went to see the same psychiatrist who had treated my last depression. She gave me a prescription for Seroxat, the same antidepressant I had been on before.

My decline continued. Were I a better writer I could, perhaps, describe it. But the feelings were so different from day-to-day life that it is very difficult. I began to have what seemed like panic attacks. A cold, tingling feeling would spread over my skin, and my anxiety would increase. There was the sensation that I would faint. My heart seemed to beat slower rather than faster, and walking around helped a bit. Taking a beta-blocker also helped. Sometimes it would last minutes, other times, hours. My cognitive therapist was very supportive and taught me breathing techniques to deal with panic. It was hard to tell what set these episodes off. When I told my psychiatrist that I was not depressed but extremely anxious, she was not persuaded – she was convinced I was having another episode of depression.

Exercise had always helped me with mild depression or when I was feeling low. I tried playing tennis once. To my partner's surprise I played extremely well, even though I was barely able to speak. I also tried jogging, but this made me exhausted and when I lay down to rest I entered into a very strange half-sleep state in which I no longer had control of my thoughts. I felt I might be going insane – a term which carries less stigma than 'mad'. My therapist and psychiatrist assured me that I would recover. I did not believe them. My experience of recovering from my previous depression was no help at all. I had learned nothing. When a friend urged me to read my own book I was not even amused.

Unlike the previous occasion I was not suicidal, but felt all the time as if I had low-level flu without a temperature. I could go to work for only a few hours, after which I felt too ill to continue. Some of this may have been side-effects of Seroxat, which included nausea. Important meetings were coming up in November and December: a major lecture in Germany, a meeting of which I was a co-organizer at the Royal Society in London, a trip to Brazil to give five lectures. The possibility of my attending any of these was becoming increasingly unlikely. It was just possible to fulfil my teaching commitments, but the idea of travelling was terrifying. What would I do if I had a severe panic attack en route? I cancelled all meetings, and even today still feel very bad at having let people down. It is one of the weird aspects of depression that it is now hard to recall the feelings that made it impossible for me to go.

There were days when I could barely get up. I thought often about the possibility of asking to be hospitalized again, and my psychiatrist arranged for this eventuality. My partner found my condition very difficult to deal with. She persuaded me, against my own wishes, to see a psychoanalyst, even though I continued to see my cognitive therapist weekly. She had to drive me there and pick me up as I was unable to travel there on my own. I would not lie on the couch, and in several of the sessions I was in a state of panic. The analyst was quite sensible and advised me to have a proper medical check-up, which I did, and was found on all tests to be normal. But this type of analytic therapy did not work, and not just because of my conviction that most of the ideas in psychoanalysis are without any scientific foundation. There were two main reasons. The first became apparent when the analyst said to me that I would not get better without his help–it seemed to me that he was trapping me for years to come, and I found the suggestion immoral and intolerable. The second reason related to his diagnosis of what was wrong with me. Nothing new: I was frightened of my impending retirement and needed other people to maintain my own identity. He also claimed I was repeating an adolescent breakdown, but I had no real memory of one. But my difficulty was, even if I granted all that was true, what was to be done about it? What could he offer? Nothing, it seemed, except asking me to lie on the couch and free-associate several times a week at £94 a session. Neither useful nor acceptable.

Fortunately he took a long – three-week – Christmas break, and I slowly began to get better. It was probably the Seroxat, but it had taken at least six weeks to have an effect. The first indication was that there were days on which I did not have a panic attack. An important positive step was going to a party at a sympathetic friend's and being able to get through the evening – this may seem trivial in retrospect, but it was a major achievement for me and increased my confidence. By Christmas I was able to spend the day with my partner and her family, and we even went to a New Year's Eve party which I enjoyed. I was getting better, and by the time my psychoanalyst returned in early January, I felt well enough to tell him I would not come to any more sessions.

My cognitive therapist was much more helpful. For me it was a much better relationship as she had none of the analyst's secrecy about himself – he would never even tell me where he was going on holiday. My relationship with my cognitive therapist was much more open and personal: we could discuss practical issues like whether and when to cancel impending commitments such as lectures, and how I might prepare for them. She encouraged me to rest as much as possible and not feel guilty about it. There was no attempt to resort to what I regard as psychobabble; rather, we discussed openly all my problems and what might be their causes.

Since then my improvement continued and I have even become slightly, just slightly, manic, possibly as a result of the

Seroxat, which I continue to take, but only a rather small dose. My psychiatrist wants me to continue for at least a year. It seems the right thing to do as the side-effects – sexual activity apart – are minimal. I may well take it for the rest of my life. There seems to be no good reason for not taking a drug such as an antidepressant for long periods if it can avoid recurrence of the illness.

I now have to face the possibility that I will have another episode of depression. When asked what actually set off my last one, I must confess I do not really know. I can make a story that it was related to my anxiety about retirement, but how do I know if that is really the explanation? More important is how to avoid another episode. All the reliable evidence is that both cognitive therapy and antidepressants can postpone if not entirely prevent another attack. And for myself and others I strongly advise getting professional help as early as possible.

(2001)

Since the last edition of this book I have had three further episodes. It really is a chronic condition. These depressive episodes were different and I was very tired and a bit unwell in the mornings but improved towards evening. I did not actually feel depressed but my psychiatrist told me that it was depression and that I was somatising – the depression was causing bodily symptoms. I did get very tired after jogging. I went back onto Seroxat and after a few months got better. But back it came a year later at the end of a holiday, this time accompanied by nausea. My new psychiatrist – my earlier one had retired – decided that I should change antidepressants. This was a very difficult period with much feeling of unwellness and tiredness, but I ended up on Efexor and eventually felt worriedly well for nearly a year. Great; but it did not last, and when I had to have urological examinations for bladder and prostate cancer which involved biopsies, it all came back again, and I had to cancel going to meetings in Portugal and New Zealand, and is still with me to some extent as I write.

The aim of this edition is to briefly bring some of the topics up to date. Of particular interest are the possible dangers of antidepressants, the discovery of genes that predispose to getting depressed, the possible role of the immune system, some views of the evolutionary biology of depression, and what could be done to improve mental health literacy. It is an enormous subject and I am aware how far there is still to go.

(2006)

CHAPTER ONE

The Experience of Depression – Past and Present

Severe depression is a weird state – if you can describe your depression you almost certainly have not truly experienced it. Until one has experienced a debilitating severe depression it is hard to understand the feelings of those who have it. Severe depression borders on being beyond description: it is not just feeling much lower than usual. It is a quite different state, a state that bears only a tangential resemblance to normal emotion. It deserves some new and special word of its own, a word that would somehow encapsulate both the pain and the conviction that no remedy will ever come. We certainly could do with a better word for this illness than one with the mere common connotation of being 'down'.

Major or severe depression, also known as clinical depression because it is disabling, should be distinguished from a milder depressed mood. For some sufferers the main feeling is an overwhelming sadness which can be accompanied by numbness, dullness and apathy: thoughts of suicide are common, as are crying spells. Yet others can become very irritable, even angry. Difficulties with sleeping are common too, as are fatigue and a lack of energy. In severe cases the patient can hardly move and is almost comatose, and may experience hallucinations and delusions. Almost always there is also an inability to concentrate for long or to make decisions. There may be a general feeling of hopelessness coupled with a loss of self-esteem. Often anxiety is the dominant emotion and this may lead to hypochondria –

excessive worries about one's health, each apparently abnormal bodily symptom being interpreted as evidence of a major illness. A characteristic feature of depression is the loss of interest or pleasure in almost all activities. Even when something good happens the depressed mood does not improve. It is also characteristic that the depression is worst in the morning and associated with early morning awakenings.

The terms melancholy and depression are closely related and melancholy is the term that was usually used to describe the condition until quite recently. But while the term depression to describe a mental condition is often thought of as having a modern origin, it was actually used in 1665 in Baker's *Chronicle*, which referred to someone having 'a great depression of spirit'. It is also used in a similar sense by Samuel Johnson in 1753, and George Eliot in *Daniel Deronda* writes, 'He found her in a state of deep depression'. Yet, as William Styron so brilliantly puts it, depression is a word 'that has slithered through the language like a slug, leaving little trace of its intrinsic malevolence and preventing by its very insipidity, a general awareness of the horrible intensity of the disease when out of control'.

The clinical features of depression are well described by one of the pioneers of its study, the German psychiatrist Emil Kraepelin, writing in 1921:

> He feels solitary, indescribably unhappy, as a "creature disinherited of fate"; he is skeptical about God, and with a certain dull submission, which shuts out every comfort and every gleam of light, he drags himself with difficulty from one day to another. Everything has become disagreeable to him; everything wearies him, company, music, travel, his professional work. Everywhere he sees only the dark side and difficulties; the people around him are not so good and unselfish as he thought; one disappointment and disillusionment follows another. Life appears to him to be aimless, he thinks that he is superfluous to the world, he cannot constrain himself any longer, the thought occurs to him to take his life without knowing why. He has a feeling as if something has cracked in him.

There is nevertheless something absurd about the depressive state, for the feelings and thoughts of the depressive can bear so little relation to reality. Some of these almost ridiculous features are described by the writer Andrew Solomon in an article for *The New Yorker*. He describes lying in bed too frightened to take a shower. While he could mentally rehearse all the steps that were required to get him to the shower, they became like 14 steps as painful and difficult as the Stations of the Cross. Even though he knew that he had effortlessly showered every day for years he now hoped that someone else would open the bathroom door. It all seemed so idiotic and hopeless, particularly as he had done skydiving, and it seemed that it had been easier to make his way toward the tip of a plane's wing against a powerful wind at 6,000 feet than it was now to get out of bed and take a shower. No wonder that he wept.

If we had a soul – and as a hardline materialist I do not believe we do – a useful metaphor for depression could be 'soul loss' due to extreme sadness. The body and mind emptied of the soul lose interest in almost everything except themselves. The idea of the wandering soul is widely accepted across numerous cultures and the adjective 'empty' is viewed across most cultures as negative. The metaphor captures the way in which we experience our own existence. Our 'soul' is our inner essence, something distinctly different from the hard material world in which we live. Lose it and we are depressed, cut off, alone.

Depression, or melancholy as it was known, has a long history, probably as long as that of *Homo sapiens* itself, and there are descriptions going back to the earliest literature. It is present in the Bible. Listen to Job's despair: 'Why is light given to those in misery, and life to the bitter of soul, to those who long for death that does not come, who search for it more than for hidden treasure, who are filled with gladness and rejoice when they reach the grave?' (Job 3:20–22)

It was, and still is, common in various cultures to attribute the cause of mental illness to a supernatural agent. In Ancient Greece it was believed that mental illness could be inflicted by the gods as a punishment for some misdeed. In early Christian times it was sometimes considered to be a test of the faithful, sent by the Devil. Melancholia as a distinct medical condition was, however, already recognised in Greece in the 4th century BC in the Hippocratic writings. It was associated with aversion to food, despondency, irritability and restlessness and fear. The leading authority on medical conditions in the 2nd century BC was Galen, whose humoral theory lasted for centuries to come. The explanation for the condition was in terms of an imbalance of the four Galenic humours – blood, yellow bile, black bile and phlegm – that were thought to govern human well-being and illness. Melancholia was thought to be due to an excess of black bile. Galen's description of the condition has a contemporary ring: 'Although each melancholic patient acts quite differently than the others, all of them seem to be filled with fear or despondency. They find fault with life and hate people but not all want to die . . . Others again will appear to you quite bizarre because they dread death and desire to die at the same time.'

It is somewhat ironic that in earlier times there was not always the stigma attached to depression that there is today, and that the melancholic thought of himself as a rather superior being. For Aristotle, melancholy was the temperament of the creative artist, for creativity was thought to be driven by black bile. Aristotle had an influence on attitudes to melancholy that lasted for centuries, since he asked why it was that those who became eminent in philosophy, politics, poetry or the arts, as well as many of the great Greek heroes, were of a melancholic temperament. He included among these Plato and Socrates. There could be, he suggested, a touch of mad genius in melancholia, and so melancholy was an enviable condition of the mind.

By the late 4th century the Christian Church was using the term to refer to 'a weariness or distress of the heart' – a condition that was regarded as undesirable and requiring treatment. While initially associated with sadness it later became associated with the 'sin of sloth' and known as accidie. Accidie in the 1300s was listed by the church as a cardinal sin for it made, for example, monks lazy and sluggish. For St Thomas Aquinas, accidie was the result of shrinking from doing some good. But the concept of accidie is more complex than that, and interpretations vary.

Some commentators related the origin of black bile to Adam's eating of the forbidden apple. With the weakening of the power of the Christian Church in the 15th and 16th centuries accidie became more and more associated with melancholia.

An Arabic medical writer in Baghdad in the early 10th century wrote a treatise on melancholia claiming that black bile was its immediate cause. His definition of the illness is striking: 'A certain feeling of dejection and isolation which forms in the soul because of something that the patients think is real but which is in fact unreal.' He describes those afflicted as 'sunk in an irrational, constant sadness and dejection, in anxiety and brooding'. He attributed mental overexertion as a major cause of the condition, but also recognised the role of bereavement and loss of possessions.

Paracelsus, a leading medical writer in the Renaissance regarded melancholy as a form of insanity. His suggestion as to how it should be cured – 'If a melancholic patient is despondent make him well again with gay medicine' – is alas, quite the wrong way to proceed. The term melancholy as used in the scientific literature of the time referred to a cold dry humour normally present in the body. This natural melancholy could be corrupted by heat and so form a noxious humour. The term melancholic could also denote a person in whom black bile was dominant and could cause physical infirmities, fear and sorrow. This condition could worsen to give rise to a mental disorder with excessive sadness and fears, lethargy and a dislike of humankind. An improper diet was often thought to be the cause. Bloodletting to eliminate the offending humour, and warm, moist air and mental diversion, were strongly recommended.

The idea of melancholy began to appear frequently in English literature in the 1580s and the word was in common use in England during the Renaissance. In contrast to the medical perception of melancholy, Aristotle's views persisted, and Robert Burton, for example, asserted that 'melancholy men are of all other the most witty'. It was thought that melancholy encouraged intellectual and creative talents. Yet Hamlet with his black clothing and lack of sociability, his morose brooding and suicidal thoughts, would also have been totally consistent with the Elizabethan conception of a melancholic man:

I have of late (but wherefore I know not) lost all my mirth, foregone all custom of exercises; and, indeed, it goes so heavily with my disposition, that this goodly frame, the earth, seems to me a sterile promontory; this most excellent canopy, the air, look you, – this brave o'erhanging firmament, this majestical roof fretted with golden fire, – why, it appears no other thing to me than a foul and pestilent congregation of vapours.

There were several treatises that could well have had an influence on Shakespeare. *A Discourse . . . of Melancholicke Disease* by du Laurens was typical and described the sadness of the melancholic and the fitfulness of their sleep; sadness without cause was a common description. In Burton's *Anatomy of Melancholy*, humoral theory remained central. His description included many physical disorders such as headache, bellyache and palpitations, and there is little reference to guilt. While he recognised grief associated with bereavement as a possible cause of melancholy he complained of the confusions and contradictions in deciding just what melancholy is. As a working definition he chose 'a kind of dotage, without a fever, having for his companions fear, and sadness, without any apparent occasion'. 'Never despair,' he counselled the melancholic. 'It may be hard to cure but not impossible for him, that is most grievously affected, if he but be willing to be helped.' He advised the use of prayers and 'physic'.

By the late 17th century the humoral explanations of Galen were giving way to chemical and mechanical ones. The latter particularly gained pre-eminence in the 18th century, influenced by a Newtonian, mechanical view of the world. Thus, for example, Harvey's discovery of the circulation of the blood led to theories which were based on a faulty circulation, and these then gave way to theories that emphasised the electrical properties of the brain. But, as in the 17th century, treatment was still largely Galenic – bloodletting, cathartics and emetics were used to drain the body of the black, melancholic humour.

In 1691 Timothy Rogers wrote a book about his own melancholy which was for him 'the worst of all Distemper; these sinking and guilty fears which it brings along with it, are inexpressibly dreadful'. He often felt that God had departed from his soul and he frequently contemplated suicide. In 1733, an Edinburgh doctor, George Cheyne, himself a depressive, wrote of the 'English Malady', by which he was referring mainly to those with a 'deep and fixed melancholy', a condition he ascribed to at least a quarter of the middle and upper classes. Another author, William Cowper, in 1773 was 'plunged into a melancholy that made me almost an infant' and he too thought of himself as 'deserted by God'. John Donne wrote in the 17th century that 'God has accompanied, and complicated almost all our bodily diseases of these times, with an extraordinary sadnesse, a predominant melancholy, a faintnesse of heart, a cheerlessness, a joylessness of spirit'; this view of melancholy persisted until late in the 18th century when there was a change in medical perceptions, and mental disorders were seen as a disorder in the brain rather than the blood or the soul.

Patterns of negative feeling are very common characteristics of depressed people. In this state, the recall of pleasant experiences is difficult. John Stuart Mill records in his autobiography the experience of such negative thoughts and the inability to enjoy anything:

> In this frame of mind it occurred to me to put the question directly to myself, 'Suppose that all your objects in life were realized; that all the changes in institutions and opinions which you are looking forward to, could be completely effected at this very instant: would this be a great joy and happiness to you?' And an irrepressible self consciousness distinctly answered, 'No!' At this my heart sank within me: the whole foundation on which my life was constructed fell down. All my happiness was to have been found in the continual pursuit of this end. The end had ceased to charm, and how could there ever again be any interest in the means? I seemed to have nothing left to live for.
>
> At first I hoped that the cloud would pass away of itself; but it did not. A night's sleep, the sovereign remedy for the smaller vexations of life, had no effect on it. I awoke to a renewed consciousness of the woeful fact. I carried it with me into all companies, into all occupations. Hardly anything

had power to cause me even a few minutes oblivion of it. For some months the cloud seemed to grow thicker and thicker. The lines in Coleridge's 'Dejection' – I was not then acquainted with them – exactly describe my case:

> A grief without a pang, void, dark and drear,
> A drowsy, stifled, unimpassioned grief,
> Which, finds no natural outlet or relief
> In word, or sigh, or tear.

In vain I sought relief from my favourite books; those memorials of past nobleness and greatness, from which I had always hitherto drawn strength and animation.

Considering how widespread depression is, there are few descriptions in the English novel. Suggestions, for example, that Lucy Snowe in Charlotte Brontë's *Villette* and Pip in Dickens' *Great Expectations* suffer from depression are misleading, for in both cases, while the characters are on occasion very unhappy, that is a long way from depression. Perhaps depression is so negative a condition that authors have avoided describing it. Nevertheless, the absence in novels is made up for by poets' and authors' descriptions of their own depressions. Gerard Manley Hopkins' poem is a disturbing description of the pain of depression:

> No worst, there is none. Pitched past pitch of grief,
> More pangs will, schooled at forepangs, wilder wring.
> Comforter, where, where is your comforting?
> Mary, mother of us, where is your relief?
> My cries heave, herds-long; huddle in a main, a chief-
> Woe, world-sorrow; on an age-old anvil wince and sing –
> Then lull, then leave off. Fury had shrieked 'No ling-
> Ering! Let me be fell: force I must be brief'.
> O the mind, mind has mountains; cliffs of fall
> Frightful, sheer, no-man-fathomed. Hold them cheap
> May who ne'er hung there. Nor does long our small
> Durance deal with that steep or deep. Here! creep,
> Wretch, under a comfort serves in a whirlwind: all
> Life death does end and each day dies with sleep.

The French poet Gérard Nerval used the metaphor of the black sun to sum up the blinding force of depression in his poem 'The Disinherited', which starts with the lines:

I am saturnine, bereft, disconsolate,
My Prince of Aquitaine whose tower has crumbled;
My lone star is dead and my bespangled lute
Bears the black sun of melancholia.

The mood of misery and suffering that usually accompanies depression was expressed by Edgar Allan Poe in a letter written when he was in his mid-twenties:

My feelings at this moment are pitiable indeed. I am suffering under a depression of spirits such as I have never felt before. I have struggled in vain against the influence of this melancholy – You will believe me when I say that I am still miserable in spite of the great improvement in my circumstances. I say you will believe me, and for this simple reason, that a man who is writing for effect does not write thus. My heart is open before you – if it be worth reading, read it. I am wretched, and know not why. Console me – for you can. But let it be quickly – or it will be too late. Write me immediately. Convince me that it is worth one's while – that it is at all necessary to live, and you will prove yourself indeed my friend. Persuade me to do what is right. I do not mean this – I do not mean that you should consider what I now write you a jest – oh pity me! for I feel that my words are incoherent – but I will recover myself. You will not fail to see that I am suffering under a depression of spirits which will [not fail to] ruin me should it be long continued.

Another account comes from the contemporary neuroscientist George Gray who had a severe depression in his fifties and describes the course of the illness in terms of the inability to anticipate future pleasant events, which he calls self-grooming.

In the early stages he begins to feel physically ill, and as the days pass his mental self-grooming decreases. At the start his optimism prevails. 'I feel ill now and unable to cope, but tomorrow I'll feel better.' When tomorrow arrives, however, and he feels slightly worse, he learns that the optimism of the previous day was unjustified. This gradual unlearning of optimism continues on for hundreds, even thousands of days, all optimistic thought abolished – for the patient has learned (correctly) that the future holds nothing but terrible suffering. Mental self-grooming has ceased and day after day a thousand and one events confirm previous pessimistic thoughts and a complete breakdown results.

These descriptions might provide some small sense of what it is like to be depressed even though severe depression is virtually indescribable to anyone who has not experienced it. But it is essentially a Western view of depression and gives no clue as to how depression is experienced in other cultures. Depression can be experienced in different ways and in some cultures physical symptoms such as headaches and stomach pains can be dominant. Such differences raise the issue as to whether depression has features common to all cultures and is indeed a single disease.

There is another aspect to the experience of depression that is of the greatest importance but very often neglected; the effect of depression on those associated with a depressed individual. There are very few descriptions of the tribulations suffered by carers. In one study one third of partners of depressive patients were themselves found to be suffering from depression. For the carer it is often extremely difficult to understand why their partner, for example, should be in this condition, particularly when there are no obvious reasons for it – after all, we are all beset at some time or another by difficult problems and it can seem that the depressed person is just not trying hard enough. But trying to push someone out of depression or to persuade them to snap out of it does not work. While marriage can help to protect against depression, as can any close and intimate relationship, depression can put a great strain on such relationships. Marital conflict and the absence of support can cause a worsening of the condition. Moreover, life with a depressed person can make a partner angry, and many have to seek psychological help to deal with the situation.

In an experimental study subjects were asked to speak on a telephone with a patient who, unknown to the subjects, was depressed. When asked about their conversations their reports were very negative. Other studies have confirmed that depressed individuals have a negative impact on those with whom they interact. When in a position of power depressives tend to exploit their position and can be uncooperative. In low power roles they tend to blame others. I recall with some guilt that long before I had my depressive experience I had employed an assistant in the laboratory on a temporary basis. She was very good at her work but at the border of a severe depression. Her effect on the group in the laboratory was so bad that they had great difficulty working not only with her but even near her and so I decided she could not continue.

The serious nature of depressive illness as experienced by sufferers is evident. I now turn to the medical viewpoint: how can it be diagnosed?

CHAPTER TWO

Defining and Diagnosing Depression

Is depression an illness? For many people it is hard to think of mental illnesses in the same way as they think about heart disease or cancer. One reason is that it is difficult to keep remembering that all our thoughts, normal and abnormal, have a biological basis, as they are the result of the activities of the nerve cells in our brains. Illness is a combination of symptoms and signs; by contrast disease is the biological and psychological cause of those symptoms. This distinction is worth keeping in mind because of the difficulties in defining depression.

Depression is classified as a mood disorder, which is a bit like saying that cancer is a disorder of cells – helpful if one knows about cells. But what is a mood? A useful approach is to see it as an emotional state that persists over a longish period of time. Most emotions are transient and are responses to external events and internal thoughts. A mood, particularly a mood disorder, may, however, last for months, as in the case of depression. Another term often used to refer to a mood disorder is 'affective disorder'; the terms are interchangeable, and emotions such as fear and sadness are referred to as affective emotions. The experience of being sad or feeling low are emotions common to all cultures. It is only when such emotions become disabling that we think of them as an illness, that is, as characterising disease, a pathological state. It is clear that to understand depression it is essential to understand the psychological and biological basis of emotions.

The medical concepts of depression and melancholia in the Western world have changed over the last 2,000 years. At one time or another, these two words have denoted a disease, a mood, a temperament or merely a feeling that lasted a short time. It was only in the 18th century that the term depression began to find a place in the study of what was commonly called melancholia, and the term melancholia covered a much wider range of emotional states than we would now consider as depression or even an illness.

Benjamin Rusk (1745–1813), sometimes called the father of American psychiatry, assigned the causes of mental disorders to the blood vessels of the brain. Yet his treatment seems extraordinarily Galenian – bloodletting, emetics, reduced diet. Up until the early 19th century what we would now call a major depression would not have been diagnosed as melancholia. Rather it would have been called 'the vapours' or 'hypochondria' or classified as some other kind of nervous disorder. Melancholy has had a long association with hypochondria. In Greek times it referred to gastric discomfort including heartburn and pain. Robert Burton spoke of 'sharp belching . . . heat in bowels . . . pain in the belly . . . unreasonable sweat all over the body'. By the end of the 19th century, hypochondria was conceived as a condition characterised by a morbid anxiety about physical health and functions. Nowadays it seems reasonable to link hypochondria with depression only in those cases in which the condition is somatised; that is, the psychological disorder is manifested in physical symptoms (this is also known as neurasthenia). This somatisation has been referred to as a metaphor for personal distress.

Depression as a term for a mental disorder characterised by a reduced emotional state emerged during the 19th century. By 1860 it appeared in medical dictionaries as 'the lowness of spirits of persons suffering under disease'. It was probably a term the medical profession preferred to melancholia, as it evoked the possibility of a physiological explanation and sounded more scientific. Krafft-Ebing's *Textbook of Psychiatry* (1879) talks of melancholia being due to 'an abnormal condition of the psychic organ dependent upon a disturbance of nutrition'. By the end of the century depression was medically defined as a 'condition characterised by a sinking of the spirits, lack of courage, or initiative, and a tendency to gloomy thoughts' or 'state of mental depression in which the misery is unreasonable'.

In the early 20th century Emil Kraepelin had great influence on the thinking about depression with his famous textbook *Clinical Psychiatry*, in which he included both what we would now call manic depression and depression without mania. An important and well-accepted distinction between unipolar depressions, which include mild and severe depression, and bipolar depressions, which are associated with periods of mania, still exists. While I focus mainly on unipolar depression in this book, it is essential to recognise the existence of manic depression, as it is regarded as a quite distinct disorder but with overlapping symptoms. As regards treatment, Kraepelin offered little more than a rest cure, advising that patients be removed from 'irritating persons' and recommending confinement to bed for some period. Distractions such as company or sight-seeing were to be avoided. Adolf Meyer, a dominant figure in American psychiatry in the first half of the 20th century, emphasised the role of a faulty psychological reaction in depressive illness. He conceived of therapy as 'service on behalf of the patient' in which the physician offers the patient a sense of security and an understanding of the patient's personal situation.

There was in the first third of the 20th century much writing and discussion about the diagnosis and classification of the various types of depression and their relationship to mania. For example, involutional melancholia was characterised by uniform depression, fear, guilt and thought disturbances. An influential idea taken from Paul Mobius in the 19th century was the distinction between endogenous and exogenous disorders: the former reflects a hereditary disposition while the latter was due to a reaction to life events. Another influential idea came from Sigmund Freud's discussion of the relationship between mourning following bereavement and his suggestion that 'melancholia is in some way related to an object loss which is withdrawn from consciousness . . . the unknown loss will result in similar internal worth to that of mourning and so is responsible for the melancholy'.

I cannot but sympathise with the difficulties those earlier scientists had in dealing with depression. There was no reliable science of the mind or brain on which they could base their understanding or treatment, and it is sobering to realise that a scientific understanding of most physical illnesses only began in the late 19th century and the application of science to help patients started as recently as the first third of the 20th century.

Even today the variety of symptoms that patients can report makes diagnosis of depression difficult. Unfortunately there is no single reliable test which would establish the diagnosis, such as, for example, the test for a bacterial infection, or easily identifiable and consistent symptoms, as in measles or mumps. There was a little while ago some hope that a test that measured a hormonal response to a stimulus would provide a diagnostic tool, but unfortunately that has not been found to be the case.

It is difficult to know when a normal fluctuation in mood becomes a depression. Do varying forms of depression form a continuum, the different diagnoses merely reflecting the severity of the disorder? In other words, should the various forms of depression be regarded as reflecting a unitary condition? For example, for many years the distinction between psychotic and neurotic depression meant little more than the distinction between severe and mild, whereas now psychotic is used to refer to severe depression characterised by delusions or hallucinations. Another distinction, which is increasingly blurred but has been widely used, based on Mobius's terminology, is made between endogenous and reactive depression; the latter being thought to be caused by life events with a negative psychological consequence, while endogenous depression is considered to have a mainly biological origin.

The chief complaints of depressed patients in Western societies are feelings of worthlessness and despair, and often include suicidal thoughts. While such psychological complaints are usual, there are often accompanying physical symptoms such as headaches, stomach complaints and rapid heartbeats. For some patients even the smallest tasks require substantial effort and they may have difficulty moving. A depressed appearance is often present: the patient's face looks sad, with the eyes and corners of the mouth turned down; the patient may show signs of crying; their whole face may be frozen in a grief-stricken pose. The patient tells of lowered self-esteem and hopelessness and is often filled with quite unrealistic thoughts in which they blame themselves for their present state. On occasion, in severe cases, there may even be delusions and hallucinations. The delusion could be that the patient is responsible for some far distant disaster. It is apparently quite common for patients with a depressive disorder to feel that they are losing their minds. They ask for help, bemoaning their fate, but nothing satisfies them. While the general impression of depressed patients is that they are withdrawn and passive, this may be misleading, as anger, hostility and irritability are also frequently observed. In one study it was found that about one third of a group of depressed outpatients had sudden spells of anger accompanied by rapid heartbeats, sweating and hot flushes. Such anger episodes were particularly associated with anxiety-related depression.

Because of the variety of symptoms and the absence of any reliable objective test, the diagnosis of depression is not always straightforward. The diagnosis is based on a psychiatrist's interview with the patient, at which the patient's full medical and psychiatric history and a list of their symptoms are obtained. These are then compared with the diagnostic criteria that have been generally agreed by the psychiatric profession. A very influential work used for the diagnosis of depression is the *Diagnostic and Statistical Manual of Mental Disorders* (Fourth Edition), always referred to as DSM-IV. This is produced by the American Psychiatric Association and is the product of a number of groups whose aim is to draw on the widest pool of reliable information relating to mental health. It provides a classification of mental disorders for use in clinical, educational and research settings. Earlier editions were influenced by psychoanalytic thinking, but in the recent editions this bias has almost entirely disappeared. The authors emphasise that the criteria are guidelines that reflect the current consensus and should be used with discretion, particularly when applied to patients with different ethnic and cultural backgrounds. Another major basis for classification is ICD-10, the World Health Organisation's International Classification of Diseases.

The different types of depression are classified in DSM-IV in terms of the number of symptoms that must be present for a particular diagnosis of depression to be made. By referring to checklists of symptoms the psychiatrist puts the patient into a particular category. An essential feature of a major depressive episode in DSM-IV is a period of at least two weeks in which there is a 'depressed mood or loss of interest or pleasure in nearly all activities'. In addition, five or more of the following symptoms must be present during that two week period:

depressed mood most of the day
diminished interest or pleasure
significant gain or loss of weight
inability to sleep or sleeping too much
reduced control over bodily movements
fatigue
feelings of worthlessness or guilt
inability to think or concentrate
thoughts of death or suicide

The borderline between a major depressive disorder and less severe but chronic depression, also called dysthymic disorder, is far from well defined. The line from feeling low through mild depression to major depression may be thought of as a lumpy continuum, and the doctor has to decide just where the patient's condition lies along this continuum. It is, however, possible that major depression is different in kind as well as degree. For dysthymia, the criteria are a depressed mood for most days over a period of at least two weeks, together with two of the following symptoms while depressed:

poor appetite or overeating
too much or too little sleep
fatigue or low energy
low self-esteem
poor concentration or inability to make decisions
feelings of hopelessness

The criteria set down by the World Health Organisation in the International Classification of Diseases (ICD-10) do not use the term 'major depressive disorder' but the terms 'depressive disorder' and 'recurrent depressive disorder'. A mild disorder requires the presence of four depressive symptoms, a medium six, and a severe at least eight. While these two classifications are not that different from those just listed, they demonstrate that there is no universally accepted way to diagnose depression or even to distinguish between severe and mild depression.

A helpful distinction often used is that between primary and secondary disorders. Although at the time of the medical examination the symptoms for each may be the same, there is evidence that the outcome and the required treatment may be different. A primary disorder refers to a condition that occurs for the first time or that recurs after periods of relative normality. By contrast, in secondary disorders there is a history of mental and other illnesses. The condition may be chronic and there are no periods of relief or of a return to a near-normal condition. The associated other illnesses can be varied; a patient can present the psychiatrist with symptoms of a primary depression, but may also present with, for example, brain damage, schizophrenia, alcoholism or drug abuse.

It is somewhat surprising that negative thoughts, hopelessness, a belief that things will not get better and anxiety, all so common among depressives, are not included as symptoms of major depression in DSM-IV. Anxiety is not listed as a characteristic of depressive disorders, anxiety disorders being treated as a quite separate category. One proposal – the unitary model – suggests that anxiety and depression are variants of the same disorder, one predominating over the other at different times as the illness progresses, severe anxiety usually preceding severe depression. Nevertheless, there is evidence, in spite of the overlap, that they may be separate disorders. For example, patients with anxiety reported social unease and maladjustment more often than depressed patients, describing themselves as poor mixers, and were more sensitive to criticism. In what is classified as Generalised Anxiety Disorder the patient has an excessive anxiety about a number of events or activities and finds it difficult to stop worrying. This is associated with disturbed sleep, difficulty in concentrating and being easily fatigued. The anxiety is always out of proportion to the feared events, which may even include money, children and job security.

Major or severe depression and mild depression or dysthymia represent the core diagnoses. But there are a whole family of related diagnoses of disorders which, while similar in key features to these two, have other special features. The term 'specifier' is used to distinguish the different members of the family and these include agitated depressions, atypical depressions and depressions due to the effects of drug abuse, like cocaine withdrawal or the effect of medication. Thus, a quite large category of mood disorders in DSM-IV are 'mood disorders not otherwise specified', such as postnatal (postpartum) depression, bereavement and seasonal depression.

Postnatal (postpartum) depression in women, which occurs shortly after the birth of a child, has features in common with major depressive disorders. Quite often there are delusions concerning the new-born infant, such as that the child is possessed by the devil, has special powers or is destined to have a terrible fate. There may even be thoughts of doing violence to the child.

Women often feel very guilty about such feelings particularly at a time when they, and all those around them, believe that they should be happy.

Among mood disorders there are also abnormal reactions to bereavement. Usually a normal reaction to bereavement persists for several months and then improves steadily without treatment. Grieving individuals can often present with emotional states and symptoms, similar to those experienced in a depressive episode, such as sadness, insomnia and poor appetite. A diagnosis of major depression is not made unless the symptoms persist after two months and particularly if these are associated with feelings of, for example, guilt, or there are hallucinations or impaired motor control.

Seasonal affective disorder (SAD) is characterised by recurrent episodes of an affective disorder with a well-defined relationship with the time of year; winter SAD, for instance, is characterised by recurrent depressive episodes during the winter. It is thought that the onset of the depressive episode is related to the decrease in sunlight. Special features of this kind of depression are lack of energy, excessive sleeping and a craving for carbohydrates and a consequent gain in weight.

Hypochondriasis, or more familiarly hypochondria, is a persistent preoccupation with the possibility of having one or more serious progressive physical illnesses such as cancer. It receives a quite distinct classification in DSM-IV. Patients show persistent concern with their health, and commonplace sensations such as 'missed' heartbeats generate severe anxiety. Reassurance that there is no underlying organic disease fails to alleviate the condition. These features are quite common in depressed patients and it can be difficult to determine whether what appears to be hypochondriasis is really a symptom of depression. Hypochondriacal patients have been found to be almost ten times more likely to have a major depression. In some cultures there is a higher tendency to refer to bodily as distinct from psychological 'pain'. Hypochondriasis is considered by some to be simply the somatisation of depression.

While somatisation due to the stigma and shame that can attach to mental illness has been emphasised in non-Western societies, its presence in Western societies may be grossly under-reported. In fact in one study of 1,000 patients visiting a medical diagnostic clinic, no organic medical illness could be detected in over two thirds of them. So while physical distress is a main reason for going to a general practitioner, the most common diagnosis is the absence of organic disorder. But one must recognise that it can be very difficult to tell the difference between symptoms that are caused by organic illness and those that are psychosomatic. It is important to determine for sure whether the physical symptoms are concealing a depressive illness or whether they are 'real', that is, due to organic pathology. It is curious that conversion disorders, made so famous by Freud, in which an emotional conflict was converted into, for example, blindness, deafness or paralysis, seem nowadays to be very rare. One possibility is that many of the cases Freud reported were indeed due to a physical disorder.

Fatigue has long been recognised as a symptom of depression and there is reason to believe that it is linked to chronic fatigue syndrome (CFS). Depression is more common in fatigued individuals than in those suffering from other physical ailments, and worsens the prognosis of the syndrome. Indeed it is hard, often, to distinguish between the syndrome and depression, though some sufferers show no signs of depression. There has been much controversy over the nature of chronic fatigue syndrome. Chronic fatigue is defined as fatigue above a tolerable level of severity which has been present for six months or more, and which cannot be explained by a physical disease. Since fatigue is an important symptom of depression – it is almost always among the five most common symptoms – it has been asked to what extent the two conditions are similar or even the same. Some studies found that most chronic fatigue patients scored high on a depression scale, though the patients are often most unwilling to accept that their disability has a psychological basis.

All the diagnoses above were first developed in relation to

adults, but depression in childhood and adolescence occurs more often than most parents or schools realise. There is now consensus from studies that children at least as young as three years old can have a major depression. Such children show a pervasive sadness and persistent sad mood and lack of pleasure, and for depression to be diagnosed this must occur every day for at least two weeks. But this alone is not sufficient, and as for adults at least four other signs must be present from this list: abnormal thoughts, or guilt or helplessness; changes in sleep patterns; altered appetite; loss of interest or energy; lowered concentration. It may not always be easy to distinguish major depression from mild disturbances of mood, such as can occur after a recent difficult experience at home, school or with friends. There are children who may be tearful and depressed-looking but without the full range of symptoms.

Symptoms and ranges of depression may vary with the age and sex of the child. For example, between the ages of five and 12, children are more likely to look sad, be agitated and have irregular patterns of sleep, while those between 12 and 16 are more likely to speak of depressive feelings like hopelessness and to have reduced concentration which affects their performance at school. Both groups report feeling that they would be better off dead, but children, as distinct from adolescents, rarely attempt suicide.

The symptoms foreshadowing a major depression usually develop over a period ranging from a few days to several weeks. They include anxiety and other typical symptoms of major depression but in a milder form. Once a major depression takes hold it, if untreated, typically lasts six months or longer; this timescale is not related to the age of the patient, though the average age of onset is the mid-twenties. In the majority of cases there is a complete return of normality so the depressive episode is self-limiting, though for the depressed patient six months is an agonisingly long time. In both the USA and Japan more than three quarters of major depressions are not treated, largely because the patients do not go to the doctor. In a small number of patients the major depression can continue for two or more years, and is then classified as chronic. For about a quarter of sufferers mild depression may continue for years. About one half of patients who have had a major depression will experience a further episode. And the more episodes the more likely it is that yet another will occur. So an individual with a history of three episodes has a 90 per cent chance of suffering a fourth.

Among the tools for assessing the severity of depression, used to determine the prevalence of depression in a population and the success of any treatment, are a variety of questionnaires which patients fill in. One that is widely used is the Beck Depression Inventory. It contains groups of statements which relate, for example, to feelings about the future, sadness, sleeping patterns, interest in sex and suicidal thoughts; the patient is asked to pick the statement in each group that best describes their feelings over the previous week. Two of the groups of statements are as follows:

0 I do not feel sad
1 I feel sad
2 I am sad all the time
3 I am so sad or unhappy that I cannot snap out of it

0 I can work about as well as before
1 It takes extra effort to get started at doing something
2 I have to push myself very hard to do anything
3 I can't do any work at all

There are 21 such questions and each statement is given a numerical score, which are then added up; a score of 16 and above is regarded as indicating a depressive disorder and 30 to 63 a severe depression.

Another widely used scale for rating depression is that of Hamilton. In this case an interviewer asks questions about the patient's thoughts and feelings and then rates the answers. The questions naturally include ones related to depressed feelings, but also to anxiety. For example, the patients would be asked whether they recently suffered from any of the following: trembling, shakiness, excessive sweating, feelings of suffocation or choking, attacks of shortness of breath, dizziness, palpitations, faintness, headaches, pain at the back of the neck, butterflies or tightness in the stomach. The answers are rated on a scale 0 to 4, with 0 reflecting absence and 4 severity, and the total is used as a measure of depression.

DEPRESSION

Now I would like to ask you about the way you have been feeling during the last month. Do you keep reasonably cheerful, or have you felt depressed or low spirited recently? How would you describe it? Moody? Downhearted? Dejected? Sad? How often? Does it come and go? Does it get better if you are with someone else? How long does it last? Have you wanted to cry? Does crying relieve it? Do you feel beyond tears? So bad it is excruciating or very painful?

0 = *Absent* or very mild or occasional. Feelings no worse than the patient's normal feelings when well.

1 = *Mild.* Persistent feelings described as moody, downhearted, dejected, or similar ways; more intense, occasional feelings; may be relieved by company, being at work.

2 = *Moderate.* Persistent or frequent feelings of depression, blueness etc.; often feels like crying, may cry occasionally; not easily relieved by company.

3 = *Marked.* More intense feelings; maybe frequent tears; more persistent throughout the waking day.

4 = *Severe.* Persistent severe feelings may be described as usually beyond tears, painful, little relief *or* extremely severe, excruciating, agonising, persistent, unrelieved feelings.

GUILT

Have you had a low impression of yourself? Have you blamed yourself for things you have done in the past or recently? Have you felt guilty about things? Have you felt you have let your friends and family down? Have you felt you are to blame for your illness? In what way? A little? A lot? Is your condition a punishment?

0 = *Absent* or very mild feelings of self-blame on borderline of normality.

1 = *Mild.* Lowered opinion of self with persisting feelings of regret about past actions which in themselves are not markedly unusual.

2 = *Moderate.* More intense or pervasive feelings of guilt or self-blame which are pathological in the rater's judgement.

3 = *Severe*. Pervasive feelings of self-blame, guilt, or worthlessness regarding many areas of patient's existence. This often leads to a feeling that the illness is a punishment for past misdeeds.

4 = *Delusions of guilt*. Incorrigible beliefs of pathological guilt, with or without hallucinations of voices emphasising guilt.

Of course such questionnaires and inventories have their limitations, for the answers and questions simplify the patients' feelings and responses. Nevertheless, they have provided valuable research tools for studying depression and can also provide an objectivity not easily obtained from a doctor's brief interview with a patient. But encapsulating and characterising a depressed person's illness remains extremely difficult. The doctor has ultimately to decide where on the depressive axis the patient lies and so decide on the appropriate treatment, and this decision should also take into account the patient's general situation, including their personal and social relationships, their economic status, and their general health.

Stanley Jackson at the end of his history of depression perceptively writes: 'However objective we may become about *depression* or about a particular *depressed person*, however carefully we may manage to identify neurophysiological and neurochemical factors in clinical depressions, someone else's depression, defined as clinical or otherwise, is ultimately going to come home to us as a fellow human being who also has needs, who also knows something about personal losses, disappointments, and failures, who also knows something about being sad and dejected, and who has some capacity for distressed response to such a distressing state. With such distress, we are at the very heart of being human.'

CHAPTER THREE

Mania

Manic depressive illness, also known as bipolar depression, is a different condition from the depressive disorders described in the previous chapter. It is characterised by pronounced changes in mood, depressive phases alternating with abnormally elevated moods. The depressive phases involve experiences similar to those with a unipolar depressive condition, but it is the mania, the other side of depression, that makes this illness fundamentally different. The writer Theodore Roethke captured the mystical merging of identities and experiences so common to the manic experience. He describes how he suddenly, one day, started to feel very good. He felt as though he knew what it was like to be a tree or a flower or a blade of grass, or even a rabbit. He walked around feeling wonderful, but as he was passing a restaurant he thought that he felt what it is to be a lion. He entered and ordered a steak, uncooked, and started to eat the raw meat. When he saw how revolted the other customers were, watching him, he began to realise that perhaps he was behaving a little strangely.

Kay Redfield Jamison, a psychiatrist herself, has brilliantly described her own manic depressive condition in *The Unquiet Mind*. 'There is a particular kind of pain, elation, loneliness, and terror involved in this kind of madness. When you're high it's tremendous. The ideas and feelings are fast and frequent like shooting stars, and you follow them until you find better and brighter ones. Shyness goes, the right words and gestures are suddenly there, the power to captivate others a felt certainty.

There are interests found in uninteresting people. Sensuality is pervasive and the desire to seduce and be seduced irresistible. Feelings of ease, intensity, power, well-being, financial omnipotence, and euphoria pervade one's marrow . . .' It was a feeling that she missed when she took lithium to deal with her mania.

On one occasion during a shopping spree in London she spent several hundred pounds on books with titles or covers that somehow caught her current fancy; books on the natural history of the mole, or 20 sundry Penguin books because she thought it would be nice if the penguins could form a colony. She once shoplifted a blouse because of her extreme impatience, unable to wait a minute longer for 'the woman-with-molasses-feet in front of me in line'.

The poet Robert Lowell wrote about one of his manic experiences and how he had an attack of pathological enthusiasm. He ran about the streets of Bloomington, Indiana, crying out against devils and homosexuals and believed he could stop cars and paralyse their forces by merely standing in the middle of the highway with his arms outspread. He thought he might be a reincarnation of the Holy Ghost. 'To have known the glory, violence and banality of such an experience is corrupting.'

Leonard Woolf noted the deteriorating quality of Virginia Woolf's thinking and speech as her mania worsened. 'She talked almost without stopping for two or three days, paying no attention to anyone in the room or anything said to her. For about a day what she said was coherent; the sentences meant something, though it was nearly all wildly insane. Then gradually it became completely incoherent, a mere jumble of dissociated words.'

One final example is provided by John Ruskin who wrote of his own experiences with what he called the conditions of spectral vision and audit belonging to certain states of brain excitement. 'I saw the stars rushing at each other – and thought the lamps of London were gliding through the night into a World Collision . . . Nothing was more notable to me through the illness than the general exaltation of the nerves of sight and hearing, and their power of making colour and sound harmonious as well as intense – with alternation of faintness and horror of course. But I learned so much about the nature of Phantasy and Phantasm – it would have been totally inconceivable to me without seeing, how the unreal and real could be mixed.'

The fundamentally dynamic nature of manic-depressive illness has been likened to a movie; the continuous and rapid changes are rather like the moving pictures on the screen of a cinema, in contrast with a still photograph. A psychiatrist, observing for the first time a manic-depressive patient undergoing one of the many changes in mood, from melancholia to euphoria or from depression to mania and back again to depression, might be reminded of the experience of entering a movie during the middle of the story. But no matter where one takes up the plot, the story tends to swing around again to the point where it started.

Descriptions of mania can be found in Ancient Greek writers. Arataeus wrote in AD 150 that mania is associated with joy, describing how sufferers may go to the market as if the victors in some contest of skill, and believe that they are experts in astronomy, philosophy or poetry. However, it was only in the 1920s that manic-depressive illness, bipolar depression, was properly recognised as a separate disorder from unipolar depression. There are also some reports of unipolar mania, that is, without depression. Unlike unipolar depressive disorders, where females are about twice as likely to be depressed as males, with bipolar depressives there is no difference in the incidence in males and females.

Mania is characterised by an elevated mood out of keeping with the subject's circumstances or ability. The patient seems cheerful and optimistic and may have an infectious gaiety. Some patients, however, are irritable and this irritability can easily turn to anger. The mood can give a sense of elation and an intense sense of well-being and be rather like drunken cheerfulness. This mood can rapidly swing back to depression, a switch from laughter to tears. It is common for the high spirits to be interrupted, often briefly, by depressive episodes. While much manic behaviour might seem benign, it can lead to grandiose behaviour – an over-confidence in their own abilities and excessive valuation of their achievements. In severe cases sufferers may have grandiose delusions and believe that they are great politicians or religious messiahs. They may hear voices confirming their view of themselves. They can also believe that they are being conspired against because of their special powers.

Their energy is prodigious and can be disruptive, making them talk excessively and at all hours, day and night. Their ideas are constantly changing, one topic rapidly replacing another – they describe their thoughts as teeming with ideas. With all this there is a reduction in sleep as well as a reduced attention span. More serious is a loss of awareness that they are behaving in an abnormal and socially unacceptable way. Some dress garishly, others become unkempt. Sexual desires are increased and behaviour may not be inhibited, so they may make overtly sexual comments to strangers. Extravagance is common – purchasing irrelevant objects and even giving away their money. Behaviour can also be aggressive when their wishes are frustrated. The episodes of mania may last from a few days to several or more months. There is a significantly increased risk of suicide. Diagnosis of mania is not always easy and the condition can be confused with schizophrenia. It is rare for patients to think that they are ill or require treatment.

The manic syndrome is nevertheless one of the most clearly defined in psychiatry. The criteria in DSM-IV are 'distinct period of elation or irritability' and three of the following:

overactivity
increased talkativeness or pressure of speech
flight of ideas or racing thoughts
inflated self-esteem or grandiosity
decreased need for sleep
distractibility
indiscreet behaviour with poor judgement
marked impairment in occupational or social function.

Mania is rare before puberty and is unknown in children

under nine years of age. First onset usually occurs before the age of 30. The great majority of patients have more than one episode confirming that bipolar disorder is a recurrent illness. The summer months are the worst. The interval from one episode to the next tends to decrease during early stages. If the patient has four or more episodes a year they are said to be in a phase of rapid cycling and this rapid cycling occurs more in females. There are rare cases of patients who oscillate from mania to depression and back again every two days, and in other cases the mood change between mania and depression, or vice versa, can occur in a few minutes. There are more manic depressives in the professional and managerial classes, and this might reflect their greater creativity due to increased energy and risk-taking.

CHAPTER FOUR

Other Cultures

To what extent is the experience of depression common in all cultures? Is it an illness that is universal, like heart disease or cancer, or do cultural influences play a key role? The experience, definition and diagnosing of depression that I have so far described has been almost entirely with reference to Western concepts. The Western vocabulary for depression with its emphasis on hopelessness, anxiety, loss of self-esteem and guilt, has its history in essentially white, male institutions and raises for medical anthropologists the question as to whether depression is a disease common to all peoples. To understand the nature of depression it is essential to explore to what extent the 'Western' experience of depression is similar to that experienced in other cultures. Since there is no clear biological marker of depression to make a diagnosis, how could a psychiatrist be sure that the condition afflicting a Yoruba tribesman in Nigeria is the same as a disorder afflicting a lawyer in New York or a fisherman in Nova Scotia? What may be endured in India may require therapy in New York and what is regarded as insane in Barbados may be accepted in Jamaica.

The emotional pathologies of the West are represented through feelings of sadness, elation, anxiety, and fear. These form the core for the diagnosis of affective disorders that include depression. Biological explanations of mental illness have taken a dominant role since the 1980s, particularly because of the success of drugs in controlling illnesses like depression and schizophrenia. This emphasis on the neurobiology of the brain has somewhat sidelined the importance of cultural factors and there has also been a strong movement to impose Western criteria and diagnosis of mental illness on diverse cultures, cultures whose traditions and modes of thought are very different from those in the West. In his critique of this phenomenon Arthur Kleinman points out that from a cross-cultural perspective the fundamental questions include how to distinguish normal from abnormal, and how a disorder is experienced, expressed and perceived. Non-western cultures appear to emphasise somatic symptoms of depression perhaps because of beliefs about the integration of body and mind.

Attitudes to mental illness in other cultures have undergone some remarkable changes. Comparison of psychiatric illnesses in different cultures had its origin with the visit of the influential psychiatrist Emil Kraepelin to Java at the turn of the century. He observed that the people of Java seldom became depressed and he interpreted this as due not to cultural factors but to the genetic makeup of the Javanese – he believed they were psychically underdeveloped and mentally degenerate compared to people in the West. This absurd racist view reached its peak with the claim by the psychiatrist J. C. Carothers that there was a striking resemblance between African thinking and that of Europeans who had been leucotomised (part of their brain had been removed). He concluded that non-European societies had no emotional disorders and the 'noble savage' who was not troubled by modern civilisation could not experience the feelings associated with depression.

It was thus held that depression was rare, or even unknown, in Africa. However, it is now recognised that there are high rates of depression in Africa. The attitude to mental illness has so changed that some medical anthropologists have suggested that the expression of mental illness, while making little sense to others, is a meaningful reaction on the part of the individual to his situation, and thus is significantly affected by the individual's cultural background. An extreme view is to deny that there is any basic similarity in depression, either biological or psychological, in different cultures.

Another way of looking at the problem is to consider how to distinguish between normal sadness, an emotion common to most people in their everyday lives, and what a psychiatrist would regard as pathological. Anthropologists who study non-Western cultures do not think of sadness, hopelessness and demoralisation as clinical disorders. For some anthropologists emotion is not based on biology but is a cultural judgement that people use to understand their situations and relationships to others. There are claims, for example, that among the Kaluli of Papua, New Guinea, there is no word for depression, that there are very few recorded cases and that the people give full and dramatic expression to sadness and grieving. It is suggested that in Sri Lanka hopelessness is positively valued as it fits with the Buddhist view of the nature of the world. In Iranian society, while sadness and grief are valued as showing personal depth, depression is recognised as a condition requiring treatment. While each culture's beliefs about normal and abnormal behaviour are distinctive, it seems that all societies have some notion of madness; but it is not clear whether depression fits into this category.

It is often very difficult to find words or phrases for depression in non-Western cultures. The closest Yoruba description seems to be 'the heart is weak' or 'the heart is not at rest'. Emphasis on the heart or other organs in relation to emotional states can also be found among Turks in Iran. The Xhosa of South Africa, when distressed, often refer to 'mbilini', which relates to discomfort, palpitations and throbbing in the stomach. A psychiatrist has described his clinical experience of depression in India where many patients spoke of 'pain in nerves', 'heat in head' or 'sorrow in the heart', none of which fits into the Western system of diagnosis for depression. Black patients appear less likely to experience depression, anxiety and instability as distinct feelings, while some Australian aborigines have an extensive set of words to describe grief and depression. Physical complaints are generally accepted as being more common among depressed patients in India, Africa and China.

One anthropologist has offered a somewhat wicked scenario to emphasise the dangers of categorising illness in one culture in terms of the categories of another culture. Imagine a psychiatrist from South Asia who is familiar with a local disorder known as 'semen loss' which leads to certain problems such as weight loss and sexual fantasies. Then imagine the psychiatrist making a list of the symptoms, translating them, badly, into English, and then training American psychiatrists in their use. There would undoubtedly be many cases of 'semen loss' now diagnosed in the USA but would they have any validity or value?

A good example of how careful one has to be in making a diagnosis is deciding whether someone experiencing hallucinations has a mental disorder. In Britain about half of bereaved people experience some hallucination of the dead person. In many traditional cultures it is normal to believe that the dead can communicate with the living; for example, American Indians commonly hear the dead calling them. While a characteristic of depression in Western cultures is guilt, in West Africa even mild depression may be associated with the sort of hallucinations that are, in the West, usually associated with schizophrenia.

Consider the Ashanti in West Africa, where anxiety is commonly expressed as self-accusation or fears of witchcraft, and the Yoruba in Nigeria, for whom dreams of witchcraft are common in anxiety states. Among the Dakota Sioux there is a depression-like syndrome which includes experience of thoughts of travelling to the dwelling place of dead relatives. An interesting example of a culture-bound disorder is an avoidant personality disorder quite common in Japan known as *taijin kyoufu*, which means 'interpersonal fear'. Sufferers avoid meeting and interacting with other people. They fear such encounters because they believe their blushing, imagined ugliness or body odour will be perceived as being offensive.

How, then, does one decide if someone is depressed in a culture not one's own? One starting point is whether or not the social performance of the individual has changed so much as to disrupt their normal life. But this, too, is difficult to determine, for in some cultures depression is not regarded as a disease or an abnormal condition. A psychiatric worker in a Ugandan village noted individuals that he thought were depressed – they complained of tiredness and unwillingness to work. The others made no comment on their withdrawal as food was plentiful. Failing to work would not be so easily accepted in our society.

In societies such as Ashanti and Yoruba, while there are a significant number of individuals who would be classified as depressed by Western criteria, the condition is seen as a natural result of the tribulations of everyday life. But is this really so different from brief depressive episodes in Western society, which, unless very severe, go undiagnosed and untreated? There is also the danger of confusing quantitative symptoms with qualitative ones. Is the presence of five symptoms as required for diagnosis in DSM-IV really more significant than the presence of a single one which may have a much greater effect?

Undoubtedly the most important studies in cross-cultural psychiatry relate to depression and somatisation, the expression of lowered mood by bodily symptoms (often referred to as psychosomatic illness or masked depression). These bodily symptoms are medically known as neurasthenia. The diagnosis of neurasthenia declined in the West in the early years of the 20th century, but at the same time it found what has been called a welcome home in China. It was readily incorporated into the traditional Chinese medical system which is based on the functional disharmony of vital organs and imbalance of vital energy; ideas which underlie treatment by acupuncture. Neurasthenia was classified into clinical subtypes such as 'dysfunction of the heart'. In the 1950s as many as 80 per cent of medical and psychiatric outpatients in China were diagnosed with neurasthenia. Then in the 1980s Kleinman concluded that the vast majority of the patients diagnosed with neurasthenia were, in reality, suffering from severe depression. Chinese psychiatrists, re-examining their patients accepted this view but found that some 20 per cent were still diagnosable as neurasthenic, with a mood disorder described as vexed rather than depressed or anxious.

Kleinman gives the example of a woman who had been harshly treated during the Cultural Revolution in China, was married to a bad-tempered husband and had had a stillbirth. She developed chronic headaches, fatigue, dizziness and ringing sounds in her ears. On interview she would unquestionably be diagnosed as depressed by Western psychiatrists but in the 1980s the Chinese doctors attributed the depression to neurasthenia – a condition they regarded as being due to a weakness caused by insufficient energy in the central nervous system. In cases such as this the somatic complaints were little alleviated by antidepressant drugs, but were helped when family and work problems were resolved. From the anthropological psychiatrist's viewpoint, experiences of illness are enmeshed in, and inseparable from, social relationships.

Kleinman concluded that depression, while universal across all cultures in psychological and biological terms, should also be seen as the relationship between an individual and the society in which they live. Thus the chronic pain and neurasthenia of Chinese patients can be seen as the social expression of depression. This somatisation of depression, the transfer of emotional to bodily pain, makes it easier in some cultures to obtain medical help. Chinese society was not one that willingly accepted personal emotional distress, believing in the importance of harmonious personal relationships. So it was much more acceptable to blame one's depression on physical symptoms. Mental illness under Communist regimes was seen as a bourgeois construct and so depression could not be recognised as a mental disorder: no wonder, then, that those with depression expressed it through somatisation.

Somatisation as a frequent feature of depression has been repeatedly reported in other non-Western societies. In countries such as Peru, India, Turkey and Iraq, stomach ache, headaches, dizziness and lack of energy are commonly at the core of neurasthenic illness associated with anxiety and depression. The attitudes of patients at a psychiatric clinic in South India show that, while psychological distress was present, more than half of the patients reported physical – somatic – symptoms as the most troubling. These physical symptoms correlated well with the patients' belief in the stigma which would be attached to admitting to psychological problems. One 25-year-old man explained the adverse impact of social stigma on the possibility of a marriage: 'No, I would not have married if I had all these problems earlier. I feel it may affect my sister's or my daughter's marriage. People say that I have these problems, like sadness.' Other reasons for this somatisation include the idea that poor, ill-educated patients lack the linguistic skills to express their emotional experiences.

Research has indeed found a positive relationship between the severity of depression and stigma in India. Thus typical comments by depressed patients were: 'I have not told this problem to anyone because they will ill-treat me'; 'If my illness continues, I myself will not agree to marry'; 'I don't want to tell anybody. Many of my neighbours are thinking less of me.' The comments of those with somatic complaints were quite different: 'I have told everything. My friends know my problems'; 'Only if I tell will they take me to a doctor'; 'What is there to hide?' All this probably reflects the fact that, while somatic symptoms are familiar to all, depressive ones are considered to be private, even dangerous.

So depression, depending on the culture and the viewpoint taken, can be a normal feeling or a symptom or a disease. Depression experienced as physical symptoms such as headaches or back pain may be different from depression experienced as despair and may not be due to the same disease; certainly the experience of the illness is not the same. But the strong possibility exists that there are processes in common, both psychological and biological. In DSM-IV depression, anxiety and somatoform disorders are separate classifications, but in developing countries where many patients present with a mixture of symptoms these distinctions may not be valid. The experience of psychiatrists in

South India leads them to wonder if, rather than referring to masked depression in patients with somatic symptoms, it might be valid to consider Western patients who are depressed as masking somatic disorders.

Because of differing attitudes to mental illness one has to be, again, cautious about interpreting the epidemiological studies in different cultures which are considered in the next chapter. For example, in some cultures, such as those of Southern Europe, there is an accepted and valued tradition of complaining about health and mental problems – it provides a sense of martyrdom. By contrast there is, in Northern Europe, a tradition that emphasises austerity and there is a stigma attached to open expression of mental problems. Reticence about intimate problems can make cross-cultural research very difficult – this is particularly true of sexual matters. Nevertheless, with care and sensitivity it is possible to do reliable research, and as Kleinman says, 'the extreme relativism of some anti-psychiatry anthropologists is as outrageously ideological as is the universalist fundamentalism of some card-carrying biological psychiatrists'. But my own view is that there is a common biological basis to depression in all cultures because emotions like sadness are universal, though the causes and expression of depression will be strongly influenced by cultural factors.

CHAPTER FIVE

Who Gets Depressed and Why?

It is natural for anyone who gets depressed to want to know the reasons. What was it that caused the depression? This is true not only for an illness like depression but for any illness. We all want to understand why we get ill. What everyone finds intolerable is that their illness is not diagnosed and that the cause is unknown. Uncertainty is hard to bear.

My own depression began, I believe, in difficulties I experienced controlling atrial fibrillation – a common and non-life-threatening arrhythmia of the heart. The drug that had worked well for several years was no longer effective and I was going into fibrillation every week or so. Atrial fibrillation means that the regular rhythm of the heart is completely lost. While this may make one feel low, it is only dangerous if one suddenly returns to normal rhythm and a blood clot has meanwhile formed, which could result in a stroke. But I am a confirmed hypochondriac and could only think of the danger of a stroke. I became increasingly anxious as Easter approached, since I was due to go on an important trip to South Africa. I had fantasies of falling seriously ill in a remote, medically primitive environment and made phone calls to a school friend there, now a cardiologist, as to how quickly he could treat me if it proved necessary.

A change in medication gave me what I can only describe as morning sickness followed by severe stomach cramps. I persuaded a colleague at my medical school to give me an X-ray and they could find nothing wrong at all. But I was deteriorating both physically and mentally and felt quite incapable of travel. I cancelled my trip at the last moment although my doctors could see no reason for me not to go. My distress at not going, and letting several groups down, increased my anxiety, and after another change of heart-drug I began to feel very weird – I can describe it in no other way. Then, quite suddenly, I was unable to sleep at all.

Sleep had never been a problem for me. Now it seemed impossible, so I began to take sleeping pills, Temazepam, which helped a bit, but only gave me a dreamless sleep that left me feeling dopey. I found it very difficult to work, or even to get out of bed. I would wake in the morning with a burning sensation all over my body, sweating. Through the pain I realised that I was having a breakdown in the old-fashioned sense of the word. I had lost my normal psychological support systems as I was unable to work or do physical exercise; I could not even ride my bicycle. I phoned a psychiatrist friend, who came round to my flat, told me that I was seriously depressed and put me on a tricyclic antidepressant. It is for me a great mistake to read about the possible side effects of any drug, for whether by coincidence or intent I get them. I became even more obsessed with my physical symptoms. I found it increasingly difficult to urinate, and within a few days I became convinced that I would not be able to urinate at night at all; I ended up begging a urologist to hospitalise me. Wisely, he did not; but I came off that antidepressant.

Because of my hypochondria, my anxiety over my arrhythmia was, all too typically, excessive. On a previous occasion some ten years earlier, I had given a close friend and neighbour who was having a heart attack the 'kiss-of-life', and while he survived long enough to get to hospital, he died. I then took on all of what I imagined were his symptoms, and kept feeling that my blood was draining away and that I was about to die. As close friends asked, where did I think it was draining to? But the panic attacks came almost daily so I went to our Professor of Medicine, who examined me. He said he could not find anything wrong but wanted me to consult another colleague in chemical pathology. That made no sense to me, for I could not see what chemical pathology had to do with my difficulties, but I went anyhow and told him my problem. He laughed and explained that he was much worse than me and several times a week went to stand in the entrance of the hospital so that when he fell the porters would find it easier to take him to casualty. I was cured; at least for a while.

In my own case I am convinced, as I was then, that my depression was due to a drug (flecainide) that I was taking to control my heart arrhythmia. There are in fact some reports that it has been associated with depression in other patients. Moreover, there is no doubt that depression can be induced by medication for some other conditions, and can therefore have a purely biological origin, for example, as the result of an abnormally elevated concentration of the hormone cortisol. But in my case it is only one explanation, and so you should be wary about accepting it. After all, many patients take the drug I was on without getting depressed.

My wife, Jill, had a different explanation for my depression. She was convinced that my depression was linked to returning to South Africa, where my father had been murdered some years ago. That did not make sense to me as I had already been back once. Of course I preferred the drug explanation to the psychological one. But the truth may include both. Nevertheless, the drug, together with what I perceived as my loss of health, were undoubtedly major factors, combined, perhaps, with a slight genetic predisposition. All this shows that anecdote is insufficient, and it is essential to look in a rigorous way at the possible causes of depression. It is sensible to think not of a single cause, but rather of the combination of those factors that make an individual vulnerable and the external events that can trigger a depressive episode.

One approach towards an understanding of depression is to examine who gets depressed and under what circumstances. Do traumatic events in early childhood, like the loss of a parent, predispose a child to depression in later life? To what extent is the way in which parents bring up their children responsible for their vulnerability to depression? Are all depressions triggered in adults by life events that are particularly upsetting, or are there some depressions that really are endogenous? Has depression been increasing over, say, the last 50 years, or does the apparent increase merely reflect increased attention and better diagnosis? These are questions whose answers must rely not on anecdote but on careful examination of large numbers of individuals, taking care to ensure that the results will stand up to scientific scrutiny. These are carefully planned epedemiological studies, and though they are expensive and hard to perform, they are essential for our understanding of depression.

The number of possible influences on depression are enormous and range from religious beliefs to seasonal changes; there are even reports that violent thunderstorms increase the admissions of patients with depression to hospital. Before we look at external events we must consider to what extent one's genetic constitution predisposes one to depression.

Genes – we have two sets, one from our mother and the other from our father – control how our cells behave both during the development of the embryo, including the brain, and throughout our lives. The genes can do this because they control which proteins are present in the cells; and it is the proteins that are the true wizards or workhorses of the cell. They enable the cell to produce energy, to grow and divide, and to perform special functions. Genes merely code for proteins; that is, they contain the information for making the protein, and are by comparison totally passive. But if a gene is faulty or absent the result will be faulty or absent protein and this can affect cells' behaviour, with very complex results. It would thus be extraordinary if any human characteristics were not in some way influenced by genes, whether the colour of eyes, sporting skills or emotional responses. For example, Huntington's Chorea is a psychiatric illness that is due to a defect in a single gene and results in very severe psychological problems. The symptoms usually begin in middle age, starting as uncontrollable movements and progressing to dementia. All that from just the malfunction of one gene out of some 30,000.

The key concept in understanding the genetics of depression is heritability which is a measure of the extent to which the vulnerability to depression is due to genes or the environment. A heritability of 50 per cent means that genes and the environment contribute equally to the vulnerability. The way to find out about genetic influences is to look at family histories and relationships and see if there is any evidence for depression being inherited in the same way as eye colour or even complex characteristics like intelligence. All the evidence strongly suggests that the heritability for depression is more than 50 per cent; this means that more than one half of an individual's vulnerability to depression is due to the genes they have.

Studies on twins provide very strong evidence for a genetic component in depression. Identical twins have the same genes, and if one twin has a severe depression there is a 50 per cent chance that the other twin will also become depressed. For non-identical twins, if one becomes depressed there is still an increased risk of the other twin also having a depressive episode; but it is only half that for identical twins, about one chance in four. For bipolar or manic depression, there is also strong genetic influence, the risk for an identical twin being more than three quarters if the other has the illness. The role of inheritance is thus clearly very strong, particularly when one realises that identical twins are not truly identical in their bodies or brains. Development is not always exactly the same even when the same genes are present, because of what one can think of as developmental noise in the system, and also because there is in the brain a certain random element – the fine connections of the billions of nerve cells are not uniquely specified by the genes. Another factor is that the environment in the womb for each twin will be different and so affect their development.

Relatives of those with depression have also been found to have a higher risk of the illness. As many as 20 per cent of the relatives of depressed patients have some sort of depressive illness, that included manic episodes. The risk to children of parents who both have depression is more than one half. If someone has a parent or sibling with major depression then he or she will have an increased risk of about twice that in the general population, but the risk is increased to about five times if the relative became depressed before the age of twenty. If the parent or sibling has manic depression the chance of developing a similar disorder is increased tenfold. For more distant relatives there is little, if any, increased risk for either depression or manic depression. Adoption studies are particularly relevant and though the data are not always consistent they show the relative importance of genes and early environment. Someone who has a parent with depression and is raised in a foster home is more likely to get the disorder than a child of 'normal' parents raised in a foster home where one of the foster parents is depressed.

To speak about a gene for depression would be misleading, as it is equivalent to saying that the brakes of a car are for accidents; it is only when they fail to function properly that accidents occur. Genes too, have a normal function, and it is only when they are faulty owing to a mutation or are absent that they could predispose an individual to depression.

It is most unlikely that a single gene is responsible for this vulnerability to depression, and since many genes will be involved it makes identification of those genes involved much more difficult. It will nevertheless be of enormous value to identify these genes, since they could provide new insights into the nature of the disease. If we knew the genes involved we would know which proteins were at fault in depressives, and this could tell us which cellular processes were involved, and could therefore suggest new treatments.

The serotonin system provides a logical source to look for candidate genes for depression, because this system is the target of selective serotonin reuptake-inhibitor drugs that are effective in treating depression, as will be discussed. The serotonin transporter protein has received particular attention because it is involved in the reuptake of serotonin at nerve junctions. A prospective-longitudinal study of a group of children was designed to find out why stressful experiences lead to depression in some people but not in others. A group of 1037 children (52 per cent male) was assessed at successive ages until they were 26. There is variation in the length of the stretch of DNA controlling whether the transporter gene is active. One region is long, one short. These control regions were found to influence the effects of stressful life events on depression. Individuals with one or two copies of the 'short' control region exhibited more depressive symptoms, diagnosable depression, and suicidality in relation to stressful life events than individuals with two copies of the 'long' one. The stressful life events were those occurring after the 21st birthday and before the 26th. This study provides very clear evidence that a specific gene can influence the individuals' response to adverse events and can predispose them to depression.

Looking at the incidence and distribution of depression in a population can provide clues as to the causes of depression and its nature. What, for example, is the distribution of depression among different age groups and the two sexes? Does mild depression often lead to severe depression? Has depression increased in recent years as is so often claimed? And what is the incidence in non-Western cultures?

One of the problems in trying to determine the prevalence of depression in a community is the reliability of the diagnosis. Most studies are carried out by well-trained lay interviewers using questionnaires, but re-examination by clinicians of those interviewed has shown that these studies are not as reliable as one might wish. Because of these difficulties it is not surprising that the reported rates for major depression vary. Typical prevalence rates, that is, the percentage of the population who are severely depressed at any one time, are around 3 per cent in the USA and Europe, and over a period of one year the rates are around 7 per cent. The percentage of the population that will have a major depressive episode during their lifetime is about 10 per cent though some studies have found rates around 15 per cent. The largest study in the USA found that the chance of someone having a major depression during their lifetime is about one in six. The percentage of the population either experiencing depression or being in close contact with a depressed individual is thus frighteningly large. The figures for manic depression are much lower. The rates for depression in the Far East are consistently less than 50 per cent of those in the West. A survey of first year students in Japan in 2000 found that 53 per cent of the students were depressed – astonishingly high.

These figures refer to adults – what about children? This is even harder to determine than for adults, but the figures available suggest that about 1 per cent of children under the age of 12 in the UK are suffering from a severe depression while the rate for adolescents, that is, children between the ages of 12 and 16, is 3 per cent. Children who get depressed have an increased likelihood of depressive episodes when they become adults. It is only relatively recently that psychiatrists have accepted that children can suffer from depression, but it seems that depressed children have about four times the risk of developing depression when they become adults. There is also a higher risk of suicide.

Diagnosing depression in children can be confusing as it very often accompanies learning and conduct disorders, so it is hard to know which is the primary process, which is cause and which effect. A particular characteristic of childhood depression is irritability. There is also the concept of 'masked depression' which includes a whole raft of symptoms from headache to truancy. It is therefore not all that surprising that estimates of the incidence of childhood depression vary from 5 per cent to an improbable 40 per cent. Nevertheless, depressive symptoms such as listlessness, unresponsiveness and eating and sleeping disorders can be observed at any age. Because of the difficulty in making a diagnosis, many children with a depressive disorder may go untreated.

Almost everyone has a low mood at some time in their lives.

Sometimes it gets close to a depressive episode but it does not stop the person functioning. So a question is whether minor depression and severe depression are closely related, with a continuous progression from mild to severe depression – or is severe depression a quite different state from mild depression? Many people, about one in ten in the USA, have some depressive symptoms that have some effect on how they live, but they are not classified as having a clinical depression since their symptoms do not match those required for the diagnosis. Such mildly depressed people do have an increased family history of severe depression and an increased risk of having, themselves, a severe depression. All the evidence is consistent with the idea that there is a continuum between mild and severe depression. My own prejudice on this issue is that, even if there is a continuum, severe depression is an experience that bears little resemblance to mild depression.

Several reports have concluded that rates of depression have increased over the last 30 years and are continuing to do so. But how reliable are such reports, and, if they are true, what are the causes? Whatever the causes, we can discount genetic factors, as the genetic pool could not possibly change so quickly. The data on which changes in the prevalence of depression is based is of relatively recent origin, most of it being generated after 1970, for unlike other behavioural problems, such as criminality or drug-use, there has been no system for monitoring the incidence of depressive conditions. The data comes from large surveys of a community in which the present and past occurrence of depressive disorder in each individual is established, not just the age of first onset. But the evidence for an increase in the prevalence of depression over recent decades should be treated with caution. It could be that the reported increases are the result of higher awareness of depression by the subjects interviewed, together with better definitions of depression. One also cannot rule out the unreliability of memory. Nevertheless, the results from a wide range of studies in different countries have found a significant increase in depressive disorders.

A striking finding in recent years that could account for the apparent increase in depression has been an increasing incidence in the younger population. The mean age for the onset of depression is now less than 30 years. The rates for depression for those born since 1955 are highest by the age of 25. Bipolar disorders occur at an earlier mean age of around 21 years. It could be that changing cultural trends such as breakdown of traditional family life and increased social mobility may be responsible. Another suggestion is that young people have higher expectations but reduced opportunities. Contrary to what is sometimes thought, depression is not an illness in which likelihood of onset increases with old age. For people over 65 about half of those who get depressed will already have had a depression when younger. About a third of the cases are probably owing to degenerative diseases affecting the ageing brain.

A large-scale study of psychiatric disorders in Hong Kong and Taiwan found that compared to Western societies, like the United States, the prevalence of depression among the Chinese population was much lower, while the number of those suffering from anxiety, particularly in Hong Kong, was higher. In Hong Kong the prevalence of major depression was about 2 per cent and of generalised anxiety about 10 per cent. A possible explanation for these differences could be that they are due to cultural factors. In Chinese and other Eastern societies the family, and family ties, plays a much greater role than in the West. In these families there is a blurring of personal boundaries and little privacy, and individual wishes are to be balanced against the common good of the family. The authority of parents is a potent force, particularly that of the father. Thus, Chinese children become acutely aware of how their behaviour is judged by other members of the family and this may predispose them to oversensitivity in personal interactions which may lead, when a crisis occurs, to an anxiety disorder. The weakening of family intimacy in the West may, for its part, predispose children to less stable personal relationships in later life and so to depression. There is a strong correlation between depression and the absence of sympathetic social support.

The trauma of the experience of war can bring about an increase in depression in a population. Displaced and refugee children are particularly at risk. This was observed in Cambodian adolescents living in refugee camps: along with depression there were increased somatic complaints and anxiety. Children in Croatia who have had severely disturbing experiences are reluctant to talk about them, and often seek help for somatic symptoms. There was almost a doubling in the incidence of depression among soldiers who were involved in the Gulf War, and there was a similar increase in the civilian population who had experienced the stress of war in Sri Lanka. There was a dramatic increase in Beirut in the Lebanon between 1950 and 1960, a period of chaotic political change.

One of the most important and consistent findings is the difference between the incidence of depression between men and women. It is generally accepted that, in the West, the prevalence of depression is about twice as high among women as among men, while the prevalence of manic depression is similar for both sexes. Suicide is, however, much less prevalent among women than men. The reasons for this are far from clear but these figures also hold across quite diverse cultures that include those of the USA, Puerto Rico, Beirut, Korea and New Zealand.

A number of explanations for this difference have been put forward. One suggestion is that the reported difference is in fact due to women seeking help for depression more readily and so more often than men. While it is true that women attend more at doctors' surgeries this is due mainly to minor illnesses. The key evidence that confirms the difference in incidence between men and women comes from studies on large communities and does not rely on attendance at a doctor's surgery. Women consistently come out with twice as much depression compared to men and such studies exclude the possibility of the result being due to women seeking help more often.

It has been suggested that women's disadvantaged social status contributes to clinical depression. Social causes have been the focus of so much research and are often grouped together under the general term 'distressing life events'. But women do not seem to experience more such disturbing life events than men, rather they react to them with greater intensity. This could have all sorts of causes, including the biological as well as the social. Having young children at home, poverty and the lack of a confidante can all make women more vulnerable. A further factor is the low social status of, for example, the housewife. Another explanation suggests females are more prone to shame than men. One factor that in women correlates with depression is bodily shame, such as a reluctance to be naked in front of others or even to look in a mirror. It has also been suggested that women, when placed in a situation that gives rise to emotional distress, respond with a style that emphasises excessive self-analysis; they often weep and talk endlessly to friends, and write in diaries about their feelings. This is in stark contrast to men who use distraction to cope with similar problems, ignoring them altogether, or working harder, playing sports or drinking alcohol.

Could biological differences provide the answer? There is no evidence for depression being linked to the X chromosome (of which women have two, while men have just one). The more likely explanation lies in hormonal differences, women being more exposed to female hormones like oestrogen; indeed, there is some evidence that women get depressive symptoms at the time of pre-menstrual tension. Postnatal depression could certainly have a hormonal basis. Depression associated with pregnancy has long been recognised. At the annual meeting of the American Medical Association in 1893 the cause was considered to be both 'excessive lactation' and 'local (pelvic) irritations acting on the central organ (brain)'. Progress in understanding has lead to a rejection of such views but little has been confidently offered to replace them. What has improved is an appreciation of how common the condition is and an awareness of some of the predisposing factors.

A figure that is both surprising and worrying is that as many as 10 per cent of pregnant women in the USA meet the diagnostic criteria for major depression. The factors that predispose or precipitate the depression include mental problems and unwanted pregnancy. Following the birth of a child there is a significant increase in depressive symptoms in the mother. The rise is most dramatic within the first three days but the increase can persist for as long as two years. These postnatal depressive conditions have been divided up into two disorders – postnatal depression and the much milder postnatal (postpartum) blues. These conditions are found in countries as different as Sweden, Brazil, Malaysia and Japan.

Postpartum depression usually begins in the second or third week following the birth and can then develop slowly over a period of weeks or months. A common feature is somatic complaints, particularly excessive fatigue. Lack of social support, poor marital relationships and poverty are all major contributing factors. Stressful life events predispose the mother to this condition as does any previous history of depression. If the newborn baby is difficult and cries a lot this can be a significant precipitating factor. And once a woman has had a postnatal depression the risk of it recurring with subsequent pregnancies is 50 per cent. There is also the slight risk of developing a postpartum psychosis, with delusions and hallucinations.

Those suffering from anxiety or depression before the birth seem to be particularly vulnerable to postnatal depression. Women can be anxious about the impending birth of their child. The absence or withdrawal of practical help increases the probability of the mother becoming depressed. It can be a serious condition, not only for the mother but for the child, as the mothers are less attuned to their infants and have more negative responses. This, in turn, can inhibit the normal intellectual development of the child.

The much milder postnatal blues, or maternity blues, can occur in over 50 per cent of women and refers to the transient period of depression experienced by many women after childbirth.

The symptoms are mostly confined to the first week and may simply be the result of hormonal changes. The condition is characterised by rapid changes in mood, tears, irritability and even euphoria. It usually disappears after a few weeks.

So why are women more vulnerable to getting depressed? The blunt answer is that no one really knows; but it is hard to believe that the phenomenon does not have a biological basis.

Depression always occurs within a social context. Relationships, work, poverty, hopes, children, parents and so on, can all play some role in the generation of a depressive episode. To say that the origin of a depression is multifactorial is merely to say that it is necessary to try and tease out the relative importance of the various influences in a person's life. A great deal of research has therefore focused on life events and on the individual's relationships and position in society.

'Life events' is a rather broad term, used to refer to stressful external changes that are rapid, even sudden, and whose time of occurrence can be given a clear and specific date. A life event involves a substantial change in the life of the individual, related, for example, to bereavement, job change or problems with a partner or children. Research to identify such life events is not easy, for it is not possible to rely simply on the depressed person's own views, nor on those of the researcher who is interviewing them; both may bias the interpretation for all sorts of reasons, and it is hard for the researcher to avoid being empathetic. In order to minimise bias, the details of the interview are best rated by other researchers. Almost all of the studies on the role of life events have been done on women.

One of the obvious triggers for depression is the stress due to a life event involving a significant loss. When placed under severe stress an individual tries to adapt, but failure to do so, or the strain of trying, can lead to a 'breakdown' like depression. The types of life events linked to depression are mostly associated with loss of some sort or another: separation from someone, or loss due to death, or loss of self-esteem. Losses which are threatening have the greatest impact particularly if they result in humiliation or the feeling of being trapped. Changes in mood – like falling out of love – do not qualify as a life event, however important they may be. With severe life events, about one third of women have been found to become depressed and this shows what a powerful influence life events has on triggering a depressive episode.

Bereavement is obviously a major traumatic life event where the cause is absolutely clear, and so it provides an important model for understanding life events and loss in general. Bereavement increases the probability of a severe depression sevenfold. The effect is most evident in the first two years following the loss. Early intense depression after the death is a strong predictor of a subsequent depressive episode.

Loss of a spouse is obviously an extremely stressful event. It is also one of the most common for women – nearly 50 per cent of American women have lost a husband by the age of 65. Sudden loss of a spouse can result in a depression that lasts many years, especially if the spouse had not been ill. Death following an illness is easier to deal with. In a study in the USA it was found that while many women were devastated by the loss of their spouse, some were apparently unscathed, and even strengthened by the experience. The investigators tried to find out what factors determine how a bereaved woman handles such a stressful event. Women with good marriages reported more depression after the loss than women with bad marriages. Surprisingly, women with previous mental health problems were better able to cope with the loss.

Is the loss of a parent by a child a major factor in increasing vulnerability to depression in later life? There is no compelling reason to believe that the death of a parent in itself, no matter how distressing at the time, is a risk factor. What really matters is how the death is handled and the parenting and care in the ensuing period. There is, however, some evidence that the early loss of her mother, even up to 12 years of age, predisposes a woman to depression. But studies on female twins have shown that while separation from parents did increase the chances of depression, the genetic contribution was more than 20 times more important than the psychological factors. What one must always ask is what substitute was provided and how adequate it was. While early permanent separation does increase the risk of depression this is most likely to be because of family discord due to the divorce, and the consequent separation from one parent. This fits with the finding that separation due to evacuation during wartime or being sent to boarding school has no detectable effect on the incidence of depression.

Apart from bereavement, it is not always easy to identify significant life events in a depressed patient's life. Memory has a variable reliability, and there may well be distortion of events, for example, the overemphasising of an event in order to make the depression more understandable. This can thus lead to exaggeration or under-reporting of life events. A further problem is to distinguish between life events that may have caused the depression and those which may have been the result of the depression – for example, losing one's job because of an inability to concentrate and cope.

It is usually only severe life events with an element of humiliation or entrapment or the finality of death that play a crucial depressogenic role. For some the life event that triggered the depression was a threat to their identity as a wife or mother about which they could do very little, and it was part of a long history of difficulties. The precipitating event could be children getting into serious difficulties. For others it was related to feeling imprisoned and the event was related to new problems concerning, for example, housing or debt. For yet others there could be the loss of a cherished idea, as in a betrayal by one's partner or a trusted friend. Job losses or financial hardship have less of an impact – but not if the job loss involves a humiliating dismissal. The effects of events can also be indirect, but nevertheless severe. For example, a woman with a cold unsupportive partner became depressed when her younger sister married a man who was just the opposite, warm and supportive; this made her see her own situation more clearly. Again, one woman who lived in very poor housing was made much more aware of her situation when she was well cared for in a comfortable local hospital when she had a routine operation. But one must be cautious in interpreting such essentially anecdotal accounts, and they raise the question as to whether there can be truly endogenous depression – a depression that appears, as it were, from nowhere, and is not provoked by some external event. With the above examples in mind it is not difficult to imagine a depression being triggered by apparently small, insignificant events.

While acute life events have been a major focus of psychological stress research, it may be that this attention is somewhat misplaced and much more attention should be given to chronic stress. The distinction between chronic and acute is that chronic stress persists for much longer periods. Many people, when asked what their most stressful period over the previous year had been, reported chronic stress more often than acute stressful events. A study of stress in depressed individuals found that chronic stress such as physical illness, poverty and marital conflict are more powerful predictors of depression than acute stress. Chronic stress can even reduce the emotional effects of acute stress.

And things are further complicated by the fact that some people seem to experience a greater number of stressful life events than others. This might reflect their reporting and assessment of what a life event is for them. Even so there is good evidence that certain individuals experience more traumatic events – such as criminal attacks, car injuries or industrial accidents – than others. Predisposing factors would include lifestyle, personality and drug and alcohol intake. There is even evidence for a genetic component to life events. Studies on twins found that about one quarter of life events experienced were due to genetic factors.

Since about a third of women have been found to become depressed following a life event that humiliates or traps them, and about one fifth following less severe life events, the question is why they and not the others get depressed? Most women clearly manage to cope with very stressful lives. There must be something about those who do get depressed that has made them vulnerable. And there is another big problem – about one third of depressive episodes are not preceded by an obvious threatening life event and such depressions are often referred to as being endogenous.

Since many life events do not, for most people, trigger a depressive episode, is there a depressive personality, a personality that predisposes the person to being depressed? The sort of personality that is described as being predisposed to depression is one in which there is excessive dependence on personal relationships and the need for reassurance and support from others together with an inability to cope with stress. But is this saying anything more than that these people are vulnerable?

The factors that contribute to vulnerability to life events are themselves quite numerous. Foremost are genetic factors about which too little is known at this stage. Direct evidence for the role of genetic factors in one's response to stressful life events comes from a study on female twins. While genetic factors play an important role in whether or not a depressive episode will occur, those who have a high genetic risk – their identical twin has been depressed – have a twofold greater chance of becoming depressed following a stressful event than those with a low genetic risk.

Factors related to the person's relationships and social conditions have been given much attention in relation to vulnerability to life events. These include lack of employment outside the home, loss of mother by death or long-term separation before age 11, negative evaluation of self and lack of social support. It is clear that these are not independent factors, but they help predict how a woman will respond to a serious life event. Poverty, while linked to depression probably acts by making interpersonal relationships worse.

Social factors play a key role in the origin of depressive disorders. While families can provide social support to deal with such stress their very intimacy can itself be a source of severe stress. Yesterday's supporters, such as a mother-in-law, can be today's stress inducers. There is also a correlation with social class; those who are at the disadvantaged end of the social scale suffer more depression. Some researchers have even claimed that depressive disorders are always a direct reflection of social environment, both present and past. The influence of life events is greatly affected by how much the individual is cared for, loved and valued – all of which come under the term social support. The providers of social support can be spouses, partners, children, parents, work-mates, neighbours, friends and relatives. Social support can provide a buffer against adverse life events and even decrease the risk of depression when there are no life events. So absence of social support, in itself, increases the risk of depression.

An upsetting example of very severe stress that demonstrates the need for social support comes from Bruno Bettelheim, who wrote about experiences, his own and others, in Nazi concentration camps. He speaks of the importance of maintaining one's self-respect and not permitting the enemy to break one's desire to survive. When all hope was lost one could no longer prevent oneself falling prey to a deep depression. And with this feeling there was a sense of being abandoned and a consequent desire for death. If there was only some slight indication that someone was deeply concerned about one's fate, survival was possible. A small sign of care might have saved those who entered into a deep depression and turned into walking corpses.

While low self-esteem increases the vulnerability to depression in the face of new difficulties, more important is the support provided by a spouse or lover. Lack of intimacy and the opportunity to confide one's problems greatly increases vulnerability. Being let down by a husband or lover is in itself a precipitating factor which can only be partly alleviated by close ties with other members of the family or friends. Thus, while a close tie can be very supportive it also carries the risk that the woman will be let down by her partner when a new difficulty arises and this makes things very much worse. There is evidence that married women can have perceptions about husbands' support that is wildly at variance with reality. The eventual recognition of this misperception, involving as it does a loss, can precipitate a depression.

I have thus far only touched upon the role of experiences in childhood in contributing to a vulnerability to depression. It is very hard to relate early influences to later life events as there are no studies, as yet, that reliably observe early childhood experiences as well as looking at the life events in the same individuals when they are adults. Recall about events in childhood may be suggestive but it is not always reliable. Nevertheless, it has often been claimed that parents have played a major role in the later depressive and other psychiatric illnesses of their children. Is bad parenting really at the core of the causes of depression? This view has been most forcibly promoted by psychoanalysts convinced that adult depressive illness recapitulates early disappointments and a lack of gratification from the mother. Their beliefs are founded on doctrine rather than hard evidence, emerging from historical reconstructions of their patients' stories when in analysis. Evidence is all-important in such situations and requires carefully controlled investigations in which all other factors are taken into account.

It was John Bowlby, a psychoanalyst himself, who moved away from accounts provided by patients in analysis to the direct observation of infants. In his development of attachment theory he thought that parents had two positive roles: to be available and responsive when wanted, and to intervene to prevent the child from getting into trouble. Bad parenting would involve the failure to respond affectionately, even the rejection of a child, unreliable and unpredictable responses, and overprotection.

In general, adults who are exposed to adversity in childhood have an increased likelihood of mental disorder. As regards depression, the most consistent evidence suggests lack of maternal care and the presence of family violence as predisposing factors. Marital discord and divorce can also contribute. In one study, the death of a parent actually reduced the risk. One of the clearest studies of the effect of parents on their children followed the offspring of depressed parents into adulthood. These children had a threefold greater risk of having a depression and the peak age of the first depressive episode ranged from 15 to 25 years. Both major depression and anxiety disorders are commoner among the children of depressed parents than among those not so afflicted, and the same is true of children whose parents have alcohol-related problems.

Marked parental rejection or neglect, violent treatment from a member of the household or sexual abuse roughly doubles the chance of a depressive episode in any one year in adult life. In less severe cases some correlation has been found between the nature of the parenting and later depression, but this relationship is neither strong nor consistent. A group of depressed women and men reported recall of unpleasant childhood experiences during depressive episodes. Young adults who have been in care when children are more likely to have depressive tendencies than those who were not, and social disadvantage increases this effect. But since the measure of the parenting is usually based on the patients' description of their own childhood, and since the essence of depression is negativity, it follows that the depressed individual will most likely report negatively about the parental care they had been given. What one needs are ongoing studies to avoid biased reporting. Most of the evidence putting the blame on dominating, overprotective or inconsistent mothering and even neglect must be treated with care as it is almost all anecdotal. There is nevertheless some evidence that parenting style can have an effect. Anxiety disorders have been found to be associated with affectionless parental care, but with depression the association was weak. Factors in adult life must not be underestimated and close affectionate ties can compensate for earlier parental deprivation.

There is a lack of information on the relationship between occupation and depression. One would like to know, for example, to what extent footballers or soldiers get depressed. One study on top sportsmen and women in Spain found a much higher incidence of depression, and the incidence among dancers also seems to be very high. There is good data on the members of one occupation: doctors. Of psychiatric illness in doctors, about 50 per cent is depression related, women doctors being particularly affected. About one third of new doctors suffer from clinical depression. By contrast, managers showed around a 6 per cent rate of depression compared to 27 per cent for general practitioners. As always, the causes for these high rates of depression are not clear, but there were positive correlations with conflict between career and personal life, and having to take sole responsibility for difficult decisions; there is the occupational strain that doctors have as to whether the right treatment has been given, and self-criticism is a major factor in depression. But again, personality differences and earlier experiences are also important. In a study of British white-collar civil servants, strong associations were found between job insecurity and minor psychiatric problems. Work is the most important factor for inequalities in depressive symptoms in men, and work and material disadvantage are equally important in explaining inequalities in depressive symptoms in women. A lack of control in the home and work environments affects depression and anxiety. Work characteristics, including skill discretion and decision authority, explain most of the differences in the incidence of depression in middle-aged British civil servants. The higher up you are, the better you do.

Is there any correlation between religious beliefs and depression? As so often with depression, the evidence is rather paradoxical. In a study of people aged over 65 in the USA, frequent churchgoers were about half as likely to be depressed, and a Dutch study found lower rates of depression among those who were involved in a religion. But there are also indications that failure to attend religious services increased the depressive symptoms of believers, and this was true for a number of religious groups, particularly Catholics. In a study in the USA it was found that Jews had a higher rate of depression than Catholics,

Protestants and other non-Jews. The male–female ratio for Jews was most unusual, there being no difference between the sexes, and there is some evidence, somewhat puzzlingly, that this is due to a low rate of alcoholism among Jews. Among Christian groups, Pentecostals had almost twice the rates for depression compared to the others. And another surprise is that the Old Order Amish in the USA, who practise absolute pacifism and lead a life free of hostility and aggression, have a threefold higher than average rate of depression.

Are patients with life-threatening or terminal illnesses depressed? About 10 per cent of patients with AIDS show depressive symptoms, but the suicide rate is as much as 50 times greater than in the general population. For patients who have suffered a heart attack, depression in the first weeks is as high as 40 per cent and those who were depressed had a risk of dying within six months, three to four times higher than patients who were not depressed. Up to one fifth of patients diagnosed with cancer have a depressive disorder, possibly because of not only uncertainty about the future, but also the loss of control. However, depression is by no means the inevitable result of a terminal illness.

Yet depression seems even worse than severe physical illness. In 1982, Dr John Horder, one-time President of the Royal College of General Practitioners, gave an interview in which he reflected on the relative agony he had experienced when on separate occasions he suffered from renal colic, a heart attack and an episode of severe depression. 'If I had to choose again, I would prefer to avoid the pain of depression. It is a surprisingly physical sensation, with a surprising resemblance to coronary pain, because it too is total. But it cannot be relieved quickly. It even threatens life. It is oneself and not part of one's machinery, a form of total paralysis of desire, hope, capacity to decide what to do, to think or to feel except pain and misery.'

Less acute illness itself can cause depression which may go undetected; there is wide variation in the prevalence of depression in physically ill patients. About 20 per cent of patients with rheumatoid arthritis had a major depression, while 80 per cent reported loss of energy and insomnia, showing that these alone are not diagnostic of depression. Patients with kidney disease on renal dialysis show a prevalence of depression of around 10 per cent, which seems typical for patients with chronic illnesses. Patients with cystic fibrosis might be expected to be more depressed than average, but the evidence for this is poor. Skin diseases are a common reason for visiting a doctor and given the perceived psychological importance of body image it is surprising that depression is not particularly high among such patients. Nevertheless, it is thought that some cases of eczema are themselves related to depression.

While euphoria is common among patients with multiple sclerosis, 10 per cent suffer from depression. There is clear association between headache, particularly migraine, and depression. Patients who have had a stroke show increased depression; about 10 per cent were depressed before the stroke occurred. There is an association between depression after a stroke and the presence of left frontal brain lesions. Traumatic brain injury, as from a blow to the head, can also result in depression. It does seem that chronic pain is linked to depression, but it is not clear which causes which.

Seasonal depression was recognized as long ago as 400 BC by Hippocrates and there is evidence that Greek and Roman physicians treated this sort of melancholy by directing sunlight at the eyes. In more recent times a doctor accompanying an Arctic expedition at the end of the last century noted the profound influence of the absence of light on both the Inuit they visited and also the members of the expedition. He described the syndrome as involving fatigue, lack of sexual desire and a generally depressed mood.

The importance of the absence of light as a cause of depression comes from the increased prevalence of SAD (seasonal affective disorder) in northern latitudes and the worsening of symptoms as the days get shorter and darker; this also gets worse when patients move north and work in rooms without sunlight. Only about 5 per cent of people in very northern latitudes experience SAD, but a much larger proportion experience milder changes in mood. While seasonal variation in the incidence of depression is well established there is no clear evidence as to which season has the highest incidence. While it is commonly thought that depression in more northern latitudes occurs most in the winter, there are studies that contradict this. Studies have, however, found that women are more vulnerable to winter depression than men and the age of onset is around 20 to 30 years. One survey in Washington found nearly 10 per cent affected by the winter but not with full-blown depression. In other surveys more than half of those interviewed reported a seasonal change in mood.

Some patients even become depressed at the same time of the year, year after year, which may or may not be related to the season. In one reported example a lady would come into a Boston hospital on the day before Thanksgiving and ask to be admitted. There was nothing obviously wrong but over the next 24 hours she entered into a severe depressive episode. Her very consistency made her an almost ideal model for trying to understand depression, if one only knew how to do the appropriate investigation. One young intern was tempted to devote his life's work to trying.

When I visited Finland early one winter, many inhabitants of Helsinki were rather depressed, but were relieved that the snow had come at last. Winter in late November has short dark days and the rain had increased the gloom. It is the arrival of snow that lightens everything up and brings some cheer. Later on it even becomes possible to ski to work. The newspapers reported that winter depression was widespread and that one third of teachers and scientists were showing signs of SAD. So my host had installed sun-like lamps in her laboratories and the results were encouraging – people were talking more. And the fashionable Café Engel in the centre of town had also installed sun lamps next to tables by the windows, though only about half of them were turned on.

One might have thought that SAD would occur much more in Finland than elsewhere in Europe, but this is not the case. Local psychiatrists assured me that the incidence of SAD was more or less the same throughout Europe, running at about 1 to 2 per cent of the population, and this included countries like Spain and Italy. However, there was a difference in the severity of the disorder. Women are found in most studies to be much more vulnerable to SAD than men, the incidence sometimes being five to ten times higher.

All this shows that there are a large number of events that can trigger a depression, many of which are related to a loss, but these events are not the sole cause. They have to be seen within the context of the social environment, the individual's past experiences and, not least, their genetic constitution. There are even depressions that may well be endogenous and for which there is no obvious trigger; depression's onset is the result of a complex network of processes. So there are many pathways to a depressive illness. Unfortunately one of these pathways often leads on to suicide.

CHAPTER SIX

Suicide

Depression can be fatal. Suicidal thoughts are all too common among depressed patients and a frighteningly large number – about one in ten severe depressives – will kill themselves; some put the figure even higher. For manic depressives the figure is between 10 and 20 per cent. When I was at my most depressed I thought of little else. Though I am a biologist, I did not know of a fail-safe way to kill myself. I did not want to suffer any pain and did not know what drug I could get hold of that would finish me off. I hoarded my sleeping pills and heart pills but was not sure they would work, and I did not want to end up even worse off, if such a thing were possible. In hospital, my room on the seventh floor had a large window that could not be opened. I imagined smashing it open with a chair but I knew I would be too frightened of heights to jump. I am also too much of a physical coward and frightened of pain to leap in front of a train. I imagined that there was a button next to my bed that if pressed would kill me painlessly, and wondered if I would really push it. When I was at my home during the day I repeatedly imagined running across the room and crashing my head through the pane in the glass door and so cutting my throat. What stopped me? My wife became very angry and said my suicide would have an intolerable effect on her and my children. However, she agreed to help me if in a year's time my condition was unchanged. Fortunately I believed her, and so began my slow recovery.

The desire to damage or kill oneself can arise from a complex mixture of factors – inability to cope, intolerable burdens, illness. Though many depressed people have suicidal thoughts, it can be quite a long way from these thoughts to an actual suicide attempt – it is startling that about 50 per cent of adolescents have suicidal thoughts almost once a week. Suicide is by no means always associated with depression; schizophrenic patients and alcoholics also have significant suicide rates, and in non-Western cultures suicide occurs for what appear to be quite different reasons and under surprising conditions.

Many writers and poets have described their suicidal feelings. Shelley, for example:

> Then would I stretch my languid frame
> Beneath the wild wood's gloomiest shade,
> And try to quench the ceaseless flame
> That on my withered vitals preyed;
> Would close mine eyes and dream I were
> On some remote and friendless plain,
> And long to leave existence there,
> If with it I might leave the pain
> That with a finger cold and lean
> Wrote madness on my withering mien.

Tolstoy passed through a suicidal crisis before his religious conversion at about the age of 50, and described it in *My Confession*. It is curious that he describes himself as both healthy and happy and yet suicidal, but that is just one of the paradoxical features of depression:

> The truth lay in this – that life had no meaning for me. Every day of life, every step in it, brought me nearer the edge of a precipice, whence I saw clearly the final ruin before me. To stop, to go back, were alike impossible; nor could I shut my eyes so as not to see the suffering that alone awaited me, the death of all in me even to annihilation. Thus I, a healthy and happy man, was brought to feel that I could live no longer, that an irresistible force was dragging me down into the grave. I do not mean that I had an intention of committing suicide. The force that drew me away from life was stronger, fuller, and concerned with far wider consequences than any mere wish; it was a force like that of my previous attachment to life, only in a contrary direction. The idea of suicide came as naturally to me as formerly that of bettering my life. It had so much attraction for me that I was compelled to

practice a species of self-deception, in order to avoid carrying it out too hastily. I was unwilling to act hastily, only because I had determined first to clear away the confusion of my thoughts, and, that once done, I could always kill myself. I was happy, yet I hid away a cord, to avoid being tempted to hang myself by it to one of the pegs between the cupboards of my study, where I undressed alone every evening, and ceased carrying a gun because it offered too easy a way of getting rid of life. I knew not what I wanted; I was afraid of life, and yet there was something I hoped for from it.

The poet Anne Sexton suffered from manic depression and committed suicide at the age of 46. Her pain is described in this remarkable poem, 'The Sickness Unto Death':

God went out of me
as if the sea dried up like sandpaper,
as if the sun became a latrine.
God went out of my fingers.
They became stone.
My body became a side of mutton
and despair roamed the slaughterhouse.

Someone brought me oranges in my despair
but I could not eat one
for God was in that orange.
I could not touch what did not belong to me.
The priest came,
he said God was even in Hitler.
I did not believe him
for if God were in Hitler
then God would be in me.
I did not hear the bird sounds.
They had left.
I did not see the speechless clouds,
I saw only the little white dish of my faith
breaking in the crater.
I kept saying:
I've got to have something to hold on to.
People gave me Bibles, crucifixes,
a yellow daisy,
but I could not touch them,
I who was a house full of bowel movement,
I who was a defaced altar,
I who wanted to crawl toward God

could not move nor eat bread.
So I ate myself,
bite by bite,
and the tears washed me,
wave after cowardly wave,
swallowing canker after canker
and Jesus stood over me looking down
and He laughed to find me gone,
and put His mouth to mine
and gave me His air.

My kindred, my brother, I said
and gave the yellow daisy
to the crazy woman in the next bed.

The Japanese writer Ryunosuke Akutagawa committed suicide in 1927 at the age of 35. In *A Fool's Life* he writes: 'I haven't the strength to go on writing. It is inexpressibly painful to live in such a frame of mind. Isn't there anyone to come and strangle me quietly in my sleep?'

About one in five of depressed individuals will try to commit suicide. It is not always easy to recognise a potential suicide. I know someone who was contemplating death by driving into a wall so making it look like an accident. Another case involved a young woman learning to ride a motorbike so she could kill herself without her family realising the true cause. Fortunately she got better. There may be a genetic component, as with identical twins if one commits suicide there is a much greater chance that the other will do so, much greater than with non-identical twins. The number of deaths from suicides in young men in the USA has been similar to those from the Vietnam War or AIDS.

Suicide is even more highly stigmatised than depression; in many places it is illegal, rendering statistics unreliable. Only a quarter of members of the United Nations provide statistics on suicide. The World Health Organisation's figures in 1994 for suicide rates per 100,000 population put Latvia, Lithuania, Estonia and Russia at the top with figures around 40; Finland around 30; Denmark, Austria and Switzerland 25; France and

Belgium 20; Germany and Sweden 15; USA, UK, Ireland and Spain around 10.

Typically, suicide rates are higher in males and older people. Suicide rates are generally lowest in married individuals and highest among those who have suffered a bereavement or separation. High-risk occupations include doctors, especially women doctors, lawyers, people in the hotel and bar trade, nurses and writers. Unemployment increases the risk. In general, many more men than women commit suicide even though depression is much more common in women. There are, however, interesting differences among ethnic groups. Black women have a particularly low rate in the USA and it is virtually unknown in black women in late middle age. Black men too have a lower suicide rate, about half that of white males. Rates in small areas of India that have been the subject of field research give figures of around 35 per 100,000 with women outnumbering men by about three to one.

In the UK suicide rates fell during both World Wars and were particularly high during the economic depression of the 1930s. There was another peak in the late 1950s but the reason is not known. Men commit suicide by using the fumes from the car exhaust in about half of cases, while women are more likely to use drugs, including antidepressants. In two thirds of cases of suicide the intention to kill themselves was expressed to a relative, and a similar number had consulted a psychiatrist in the previous month.

Officially there are about 32,000 suicides a year in the United States, but the real number is probably much higher. It has been estimated that for each completed suicide there are 100 attempts, which gives a figure of several million suicidal events each year. The number of young people attempting suicide has been increasing. At least five people are influenced by each case of suicidal behaviour and thus the impact on society is very large. The devastation caused to families must not be, and is not, underestimated. Each year in the USA at least 60,000 children experience the death of a relative by suicide. This is obviously a very stressful life event. Adolescents who experience such a suicidal death have an increased risk of depression especially within the first year after the event.

What drives an individual to take their own life? The causes are complex and involve both biological and psychosocial factors. Psychological autopsies, in which information is obtained from the victims' relatives, friends and medical records, have established that suicidal people usually have a serious psychiatric disorder, most often depression. For depressed patients who commit suicide the most likely time is at the beginning or end of an episode. A sense of hopelessness seems to be a key factor rather than the degree of depression. Styron describes it well, writing of depression leading to a despair that 'owing to some evil trick played upon the sick brain by the inhabiting psyche, comes to resemble the diabolical discomfort of being imprisoned in a fiercely overheated room. And because there is no escape from this smothering confinement, it is natural that the victim begins to think ceaselessly of oblivion.'

A large number of alcoholics have symptoms of depression. As many as 25 per cent of drug and alcohol abusers eventually kill themselves, but it is hard to determine whether the depression causes the substance abuse or vice versa – the depression and the drug clearly interact. A study of 67 teenage suicides in the United States found that 40 per cent suffered from major depression and 30 per cent were addicted to alcohol or other drugs. Similar results were found in a Finnish study. Separation from parents and parental violence or alcohol abuse were common among the males, but it is not only in unstable or abusive families that adolescents kill themselves. Suicide does however have a moderate tendency to run in families, and individual psychological factors undoubtedly predispose some to suicide.

A famous study of suicide by the sociologist Emile Durkheim, writing before Freud, proposed that there were three types of suicide. Egoistic suicides resulted from a failure to integrate into society, throwing the individual onto their own resources. By contrast, altruistic suicide occurred when individuals were so completely absorbed in the goals of the group that they were willing to die for the sake of the group, to die for a good cause. The third type, anomic suicide, resulted from what we would now call a life event, such as a bereavement.

In his suicide note the American artist Ralph Barton predicted that those he left behind would be tempted to read into his death all manner of explanation while constantly overlooking the crucial one, his mental illness. He thought that everyone who had known him and who heard of his death would have a different explanation as to why he did it. Most of these explanations would be dramatic – and completely wrong. The reasons for suicide are illnesses of the mind. Difficulties in life merely precipitate the event, and the true suicidal manufactures their own difficulties. Barton claimed to have had few real problems and indeed enjoyed an exceptionally glamorous life. But while he had always enjoyed excellent physical health, since early childhood he had suffered from depression and in the years before his suicide he showed clear symptoms of manic depression. It prevented him from working properly and made it impossible for him to enjoy the simple pleasures of life. He rushed through several marriages and countries in what he later regarded as a ridiculous effort to escape from himself. In the end it seems that he saw only one way out.

Anthropological work has found that some individuals commit suicide who do not show signs of mental illness but are under great social pressures, and suggests that this may be a culturally acceptable way of expressing or dealing with their distress. While it is widely thought that suicide is associated with urban life and industrialisation, the truth is much more complex, and cultural differences play a very important role. In rural China, for example, suicide among young women is particularly high. The high suicide rate of some American Indians is associated with high rates of alcoholism.

China has one of the highest suicide rates in the world; some estimates put it at three times that in the West. A distinctive feature is that there are more suicides among women than men, particularly in the countryside, where the ratio is four to one.

There is even a regular feature in a popular Chinese monthly entitled, 'Why do they choose suicide?' While in the West suicide is thought to be associated with severe depression, the work of Michael Phillips in China suggests that less than half of those who commit suicide are suffering from depression as conventionally conceived. Instead it is intolerable social stress that leads to what has been termed 'rational suicide'. In the countryside, where women frequently suffer from cruel in-laws, unfaithful husbands and poverty, they kill themselves impulsively, usually by drinking insecticide, which is easily available; in the absence of adequate medical services they die.

In South Pacific island populations there has been an epidemic of suicides, and it is the leading cause of death for men. The local explanation of the high rate of suicides among young men is that it is an expression of social withdrawal and abasement because some great offence has been taken. This may be related to the elimination of traditional men's organisations and the resulting increase in conflicts within the family. It is seen as both vengeful and conciliatory. For women in New Guinea suicide is seen as calculated revenge in response to abuse, usually within a marriage. Sri Lanka has recently acquired one of the world's highest suicide rates – 47 per 100,000 in 1991, with most victims under 30 years old. The blame is placed on rapid changes in society and high unemployment. The easy access to insecticides provides a convenient means for those young people who feel hopeless and dispossessed.

A study of suicide in India found specific causes included dreaded diseases (like leprosy), poverty, love affairs, loss of money and failure in an examination. These causes are commonly blamed in all societies and emphasise the role of social factors. A major Indian medical system, the Ayurveda, does not identify suicide as a significant clinical event. Moreover, both the Jain and the Hindu cultures identify life circumstances that may be sufficient justification for suicide, though they condemn suicide in general. The legitimate grounds for suicide include a terminal incapacitating illness. Suicide by widows (*sati*) has characterised the ideal Hindu wife. The widow classically throws herself onto the pyre of the dead husband. Sati is recorded as long ago as 300 BC and it is not clear how many of those who died in this way were encouraged by relatives. There are also reports of women killing themselves in large numbers to avoid capture by invaders in the 16th century. Suicide in India was also used in the 20th century as a form of political protest.

There is a quite common impression that suicide in Japan has traditionally been accepted and even welcomed. While there was a high suicide rate in the 1950s, it has been decreasing steadily since then and is now no higher than in many European countries. It is probably the overemphasis on *hara-kiri* (ceremonial Japanese suicide) that has promoted this myth about suicide in Japan. While contemporary Japanese suicides often consider that death is the only way of resolving a desperate situation – and this is true of many other cultures – it is neither regarded as honourable nor a condoned tradition. There is a strong stigma attached to suicide and a prejudice against mental disorders. Those left behind after a suicide try to behave as if nothing has happened. The causes of suicide in Japan, as elsewhere, are complex and multifactorial. But a suicide cluster was triggered by the suicide of a famous singer Yukiko Okada who jumped to her death in 1986. More than 30 youngsters took their own lives within two weeks of her death.

For psychoanalysts, as for others, theorising the explanations for suicide has always been a problem, and Freud tackled it in *Mourning and Melancholia*:

> So immense is the ego's self-love, which we have come to recognise as the primal state from which instinctual life proceeds, and so vast is the amount of narcissistic libido which we see liberated in the fear that emerges as a threat to life, that we cannot conceive how that ego can consent to its own destruction. We have long known, it is true, that no neurotic harbours thoughts of suicide which he has not turned back upon himself from murderous impulses against others, but we have never been able to explain what interplay of forces can carry such a purpose through to execution. The analysis of melancholia now shows that the ego can kill itself only if, owing to the return of the object-cathexis, it can treat itself as an object – if

it is able to direct against itself the hostility which relates to an object and which represents the ego's original reaction to objects in the external world. Thus in regression from narcissistic object choice the object has, it is true, been got rid of, but it has nevertheless proved more powerful than the ego itself. In the two most opposed situations of being most intensely in love and of suicide the ego is overwhelmed by the object, though in totally different ways.

But is it really reasonable to accept that self-love of the ego is a primal state and that suicidal thoughts always reflect murderous thoughts towards others? What evidence is there to support such views? It is time to look at the psychological basis of depression.

CHAPTER SEVEN

Emotion, Evolution and Malignant Sadness

There is a wonderful Durer engraving in which a female figure sits somewhat dishevelled and disconsolate staring at the word 'Melancholia' emblazoned on the horizon. I am not surprised at her state. Depression is depressingly difficult to understand. People's minds are complex and their behaviour will reflect both their biology, including a genetic component, and also their past and present experiences. All these interact with each other in ways that are so poorly understood that the temptation is to throw up one's hands in uncomprehending despair. But that is too pessimistic and, while one should not underestimate the magnitude of the problem, it is possible to tease out certain factors that play a key role in initiating and maintaining and curing depression.

It may be useful to compare the problem with that of cancer, where much has been learned in recent years. A patient with advanced cancer can have numerous physical symptoms since the cancer may affect many different organs, including the brain, liver, bone and lung. Looking at such a patient before there was any understanding of the cellular basis of cancer would have been bewildering. It was only with extensive research during the 20th century that the cellular and molecular basis of cancer became clearer. We now know that cancer is the result of normal growth and cell multiplication going wrong. For example, the cells lining the human gut are continually being replaced and this requires continual cell multiplication. Which cells will multiply and how often is under very strict control, but it is inevitable that these control processes can fail and so the cells divide too often, acquiring abnormal properties that make them malignant and lethal.

Cancer originates from a single cell that has some small defect and proceeds through a series of stages from the mildly abnormal to the full-blown malignant stage. Starting with changes in genes that control cell behaviour in relation to cell multiplication and cell differentiation, the cell's offspring accumulate more and more errors. The malignant cells can then migrate from their site of origin to other organs and so can have complex effects, including hormonal ones, on the patient. But the key to understanding cancer is to recognise that it is normal control processes that have become disordered, and so, in the same way, it is necessary to understand the normal processes underlying the disorder of depression.

If we are to understand depression then we need to understand emotion. For depression is, I believe, sadness that has become pathological. Depression is a disorder of emotion. A possible common feature among all the very different emotional states is that they represent some kind of response to signals associated with reward or harm. For example, fear is a form of emotional reaction to any form of stimulus or action which the person will wish to avoid or escape from. In animals particular stimuli that are threatening can result in the response of either fight or flight. The former can result in the adoption of an upright posture with raised forepaws rather like a boxer; flight can result in jumping backwards and squealing. These behaviours can be elicited or eliminated by stimulation and lesions in specific regions of the brain.

There are a number of basic emotions which include happiness, fear, anger, sadness, disgust and surprise, and each is characterised by being initiated by a distinctive signal of rapid onset, short duration and unbidden occurrence. It is these characteristics that help separate emotion from mood. Moods such as contentment, or feeling low over longish periods, are not considered to be basic emotions as they are not initiated by distinctive signals. Depending on its duration, sadness can be either an emotion or a mood.

There is in fact good evidence for the existence of basic emotions that can be recognised by people in quite different cultures. Paul Ekman showed photographs of people expressing emotions such as happiness, sadness, anger, fear and disgust to people of different cultures, including the isolated people of Papua, New Guinea, and all had no difficulty making up an appropriate story to fit the emotion in the photograph. That these emotions are so readily recognised supports the idea that they are universal.

Evolutionary thinking tries to account for the adaptive nature of human characteristics. The obvious ones, like having hands and brains so that objects can be finely manipulated, require no novel explanation; the genes that control these abilities would obviously have been selected. Problems arise, however, as to why natural selection has not, for example, eliminated those genes that make us susceptible to disease, and further examination usually reveals that those genes have other valuable adaptive functions. A perfect example is sickle cell anaemia, a genetic disease which is due to a mutation in the haemoglobin molecule and can result in the red blood cells taking on a sickle-like shape that makes their circulation difficult, so causing serious circulatory problems. It turns out that the disease is only a problem if both of the genes coding for haemoglobin carry the sickle mutation; if only one does, the individual is much more resistant to malaria. There is thus both a risk and a benefit in having a sickle cell mutation.

Recent thinking in evolutionary psychology has tried to understand the adaptive nature of a variety of human emotions and behaviours such as sexual attraction, fear and disgust. One significant achievement has been the finding that morning sickness in pregnant women may be adaptive because the vomiting and nausea can protect the foetus from toxins. Steven Pinker views the adaptive function of emotions within his scheme of cognitive psychology in which the mind is essentially a machine for processing information. Emotions thus become an adaptation of the mind to promote the individual's survival. The role of emotions is to set the mind's goals at a high level. So fear sets the goal of avoiding impending harm, and disgust prompts the avoidance of dangerous substances. What then is the adaptive function of happiness? It is most likely that we pursue happiness because when we achieve it we are fitter and healthier, and feel comfortable and secure. It is clearly an adaptive state worth striving for. Indeed there is evidence that about 80 per cent of people in the industrialised world are fairly satisfied and about 30 per cent actually 'happy'. Yet happiness is a fragile state: not only can envy undermine it but, more importantly, there are more negative emotions, like fear and sadness, than positive ones. This means that losses have a bigger impact than gains.

In a study of the way normal people describe depression it turned out that, in spite of all the complaints about how unsatisfactory the word 'depression' is for describing the illness, the two terms mostly closely linked to depression were grief and sadness. Thus, understanding sadness is fundamental to an understanding of depression. Sadness is a universal emotion and results in a special, universally recognised facial expression. The inner corners of the brow are drawn together and the eyes slightly narrowed. Sometimes the chin is pushed forwards and quivers. Sadness is often accompanied by crying or sobbing – all of these are ways of communicating the emotion to others and are essentially a plea for help.

What then is the function of sadness? Sadness is generally considered to be a negative emotion but this in no way implies that it does not play a crucial role in our lives. The answer seems to lie in attachment. It is attachment that is adaptive, the need for the child to maintain a close bond with its mother, or carer, and for the mother to be attached to her child to help ensure the child's survival and that of her own genes. Attachment also promotes survival in a reproductive couple. For attachment to be effective the loss or removal of the mother from the child must necessarily cause distress for the child and so cause the child to search for the mother. In a world without sadness what would there be to encourage attachment to our children or our partners? What would keep an individual from breaking the ties that bind one person to another if there were no sadness at parting? Separation, whether physical or psychological, is a basic cause of human sadness.

Sadness can also result from failure to achieve an important goal. A crucial feature of sadness is that like all emotions it serves to direct one's actions in an adaptive manner, that is, it directs actions to make the survival of the individual more likely. Sadness motivates an individual to renew and strengthen personal bonds and make good certain losses, but in bereavement that is not possible and grief is the inevitable result. It may not be unreasonable to think of sadness as providing a driving force to restore attachment or loss in the same way that hunger makes us eat. Eating is pleasurable, but hunger is not.

This is essentially an evolutionary explanation of emotions and leads to the question as to why depression exists. Has depression, like sadness, some adaptive function in our evolution so that those who become depressed survive better than those who do not? A major attempt to understand depression in terms of evolution is based on the claim that depressive states represent a psychobiological response which is part of our evolutionary inheritance and must have performed some adaptive role. Given that an individual's performance is limited when depressed, it is in this impairment that they claim to see a biological function in relation to social behaviour, and especially social competition.

The social competition hypothesis for the evolutionary origin of human depression is that depression is associated with a losing strategy and, by contrast, elation with winning. Depression is viewed as an adaptation whose function is to inhibit aggressive behaviour to rivals and superiors when one's status is low. Humans, it is claimed, share with other social animals like apes a mechanism for yielding when they are in competition for sexual partners or food. Yielding prevents the aggressor from being more aggressive and promotes behaviour which encourages acceptance of a subordinate position. The hypothesis is that depression evolved out of mechanisms mediating withdrawal in dangerous social situations in order to decrease the likelihood of an attack by a dominant member of the group. This results in a social hierarchy which regulates social relationships. The 'depressive strategy' is suggested to have three main functions: to prevent the individual from being aggressive to superiors, by creating a sense of incapacity; to signal that the individual is no threat to rivals; and to put the individual in a state of mind which is accepting of their subordinate situation. Depression is thus hypothesised as 'a ritual (psychological) substitute for physical damage which is suffered by the loser in an unritualised contest'. There is evidence from vervet monkeys that when high ranking males lose their position in the hierarchy they huddle and rock and refuse food, looking rather like depressed humans. Moreover, their serotonin levels fall – low serotonin is linked to the depressive state.

I find this explanation unsatisfactory. It is based on the assumption that depression is adaptive. There are no grounds for believing that simply because it is widespread it serves a purpose, anymore than one would claim heart disease or cancer to be adaptive. Quite the contrary. Everything we know about severe depression in humans is that it is an illness; it is pathological and prevents an affected individual from functioning properly. I also cannot accept that humans are in a social hierarchy similar to that of the apes where it is a social necessity to withdraw for safety – or even that such a withdrawal is similar to depression. A more satisfactory explanation for low mood, but not for depression, is the social navigation hypothesis which attempts to account for all intensities of depression based on standard evolutionary theories. Depression, it claims, may play two complementary roles in dealing with particularly important and troublesome social problems by focusing limited cognitive resources on planning ways out of complex social problems, and by motivating close social partners, especially family, to provide problem-solving help and concessions, particularly in cases where they are initially reluctant to do so. Depression could get its evolutionary power by virtue of the costs it imposes on the depressive, and on close social partners who have a positive interest in the normal functioning of the depressive. For example, fatigue so common in depressives, motivates energy conservation and reduces goal pursuit when future effort is unlikely to pay off or when the environment is generally less propitious, such as might have occurred during the winter among human ancestors. However I would again emphasise that all this is fine for low mood, but not for debilitating depression.

To understand depression in evolutionary terms we need to understand what normal process has become perturbed and so pathological. The key lies in sadness. Any theory must account for why bereavement results in feelings and states so similar to those of depression, as Freud emphasised. The social hierarchy model fails again in this respect. If one accepts that sadness is an adaptive, universal and normal emotion then there is no difficulty in thinking of depression as pathological sadness. Something has gone wrong with the processes underlying and controlling the feelings associated with sadness, so that depression occurs which is non-adaptive. The feelings of sadness are no longer under normal controls and can persist even when the original stimuli that initiated them are no longer present. An analogy, as I suggested earlier, can be drawn with cancer, which is normal growth process going out of control; depression is sadness out of control. Thus a useful and possibly fruitful way of thinking about depression is in terms of malignant sadness. Sadness is to depression what normal growth is to cancer.

There is another feeling closely associated with depression, anxiety. Anxiety is generally experienced as an unpleasant feeling of foreboding that something bad will soon happen, and is related to fear. Anxiety is thus present in situations that threaten an individual's well-being. There are many situations that can provoke anxiety, such as conflict, frustration, physical threat and threats to self-esteem. Anxiety is a normal response to threatening situations and keeps the individual prepared to deal with them; it makes a person more alert to danger. Anxiety only becomes pathological when it is so frequent and persistent that it interferes with normal activities. Using the same line of thinking, anxiety can become malignant when the individual constantly anticipates frightening and harmful events.

The adjective malignant is very appropriate in relation to abnormal sadness, anxiety and fear as these conditions can, just like malignant cancer cells, affect – almost invade – other mental processes, and so can influence in a profound way the thinking of the affected person. They can also affect each other.

What, though, might the biological origin of mania be? One possibility is that it is related to happiness, which is clearly adaptive as it encourages one to do things that are rewarding. Thus mania could be malignant happiness. It may also be related to creativity. The existence of a link between madness and genius is an idea both old and controversial. If it were true, would it not then be possible to see, perhaps, a positive virtue in depressive illnesses? It is just such questions that Kay Redfield Jamison has explored in depth in her book *Touched with Fire*.

Aristotle praised melancholy since he thought that all those outstanding in philosophy, poetry and the arts were melancholic. It was an ancient Greek view that, as Socrates put it, 'madness, provided it comes as the gift of heaven, is the channel by which we receive the greatest blessings', for it was thought that inspiration was only obtainable in particular states of mind, of which madness was one. The Greeks recognised the distinction between those mental illnesses which were detrimental to artistic achievement and the 'sacred' madness of inspiration. There was also a fashion for melancholy in the Renaissance.

But to what extent is there really a relationship between creativity and psychiatric illness? Literary scholars have considered this question and Lionel Trilling has commented that the idea of the artist as mentally ill is 'one of the characteristic notions of our culture'. Like the Greeks, he emphasises that health is also essential for creative work, as is discipline and sustained effort. Note that here, and in most of what follows, the discussion of creativity and genius is confined almost entirely to the arts – writers, poets, painters – as if only they are creative. The relation of creativity in science, business and politics, for example, is barely touched upon. Yet it is hard to take seriously the idea that creativity is only associated with the work of those in the arts.

Anecdotal stories about creativity and mental illness will not suffice. This has been recognised and a number of scholars have tried to find direct evidence of a link. A study of the biographies of a wide range of modern artists found that the highest rate of mental illness was in poets; almost 20 per cent of those studied had committed suicide. Composers also showed high rates of mental illness, particularly depression. In general the mood disorders of artists were at least three times greater than among other professionals, including scientists and businessmen.

Jamison investigated the mood disorders of poets in Britain and Ireland born in the hundred years from 1705 to 1805. Within this group are famous figures like Lord Byron, Samuel Johnson, William Blake, William Wordsworth, Percy Bysshe Shelley and Samuel Taylor Coleridge. There was a strikingly high rate of mood disorders. These poets were, compared to the general population, 30 times more likely to suffer from manic depression and five times more likely to commit suicide. Studies on a group of 30 modern writers confirm these results in general. Almost 80 per cent of this sample had a mood disorder.

Jamison interviewed a group of nearly 50 British writers and artists, all of whom had won at least one prize or award that recognised their distinction in their field. Most were men and the average age was 53. What she wanted to know was the role of moods in their creativity. More than a third had been treated for a mood disorder and most of these had been for manic depression or depression. The artists and writers had suffered from their depression but not their mania; virtually all said that they had experienced intense, highly productive creative episodes. Almost all reported a reduced need for sleep just prior to these intensely creative periods. There were reports of mood change with descriptions of ecstasy and elation; one reported a 'fever to write'. Others experienced feelings of anxiety. For most, these moods were integral to their work.

It seems that artistic creativity can benefit from a variety of experiences, including visions, fears and melancholy. Perhaps the struggle to come to terms with emotional extremes supports the creative endeavour. Profound depression can change an individual's beliefs about the nature and meaning of life. Writers, artists and composers have described how, having struggled with a depressive episode and come through it, they have used the experience in their work. The poet Anne Sexton used pain in her work and said that 'creative people must not avoid the pain they get dealt'.

While it seems easy to understand why mania, with its surge in energy and confidence, should help with creativity, it may be less easy to see why depression could be helpful. As just mentioned, depression might put into perspective thoughts and feelings that had been generated in a more manic phase. It might serve, as Jamison puts it, a critical editorial role. Depression can force one to look inward and ask very difficult questions about oneself: what is it all about, what is the purpose, and who and what am I? Herbert Melville wrote that, 'In these flashing revelations of grief's wonderful fire, we see all things as they are; and though when the electric element is gone, the shadows once more descend, and the false outlines of objects again return; yet not with their former power to deceive.' Familiarity with sadness is important for many artists. The poet Antonin Artaud exaggerates but makes the point: 'No one has ever written, painted, sculpted, modelled, built or invented except literally to get out of hell.'

So manic-depressive illness can apparently give a touch of fire to an individual's work. Since it has such a strong genetic basis could it also perhaps be, at times, an adaptive characteristic in the evolutionary sense? It seems implausible, particularly since manic depressives are so prone to suicide and the disease is so disabling.

Most discussion of the negative effects of the mind focuses on illness. But there is another area where a relationship of the opposite kind is very clear: sport. At the highest levels of any sport – running, tennis, golf – there is little difference between the technical skills of the competitors. The way to success and winning lies in their mental attitude. How often have potential tennis champions let the game slip from their grasp by making unforced errors quite out of character with their abilities? No amateur sportsperson can be unaware of how emotions can affect performance. There is now some consensus among sports psychologists and the competitors themselves that it is necessary to maintain confidence and acquire an inner calm, to banish anxiety, tension and any negative thoughts. There is even evidence that in so clearing the mind and focusing on the sporting activity, the electrical activity in the brain becomes similar to that observed during meditation and so-called alpha waves predominate. Moreover this is correlated with improved performance. The mental attitudes that lead to sporting success are nowhere to be found in individuals with depression, anxiety or mania.

While the concept of malignant emotions has thus far been based on evolutionary considerations, it is now possible to explore how sadness can become malignant by looking at what is known about the psychological and biological basis of depression, and how sadness can invade other mental processes, altering the very way we think.

CHAPTER EIGHT

Psychological Explanations

It is always desirable in considering mental illnesses to try to link biological, psychological and sociological factors, but in this chapter I am focusing on the psychological aspects, and only in the next the underlying physiology and the processes occurring in the brain. I realise that such a separation carries a risk and that one must try to avoid thinking about brainless minds or mindless brains. My main aim is to try and understand how sadness might become so abnormal that it leads to depression. I start with psychoanalytic ideas, not because they are necessarily correct, but because they have, since the early years of the 20th century, been very influential, and two figures whose ideas have been seminal, Aaron Beck and John Bowlby, were originally analysts.

According to classic psychoanalytic theory as developed by Freud, Abraham and Klein, depression, like mourning, conceals aggression towards a lost person or object. Ambivalence towards the loss is central to the theory. As the analyst Julia Kristeva puts it: 'I love that object, but even more so I hate it; because I love it, and in order not to lose it, I embed it in myself; but because I hate it that other within myself is a bad self, I am bad, I am non-existent, I shall kill myself.' It is poetic but very hard to take as a serious theory.

The origin of psychoanalytic ideas about depression lies in Freud's classic paper *Mourning and Melancholia*, in which he compared depression with bereavement and emphasised the self-reproach and loss of self-esteem of depressives:

> The distinguishing mental features of melancholia are a profoundly painful dejection, cessation of interest in the outside world, loss of the capacity to love, inhibition of all activity, and a lowering of the self-regarding feelings to a degree that finds utterance in self-reproaches and self-revilings, and culminates in a delusional expectation of punishment. This picture becomes a little more intelligible when we consider that, with one exception, the same traits are met with in mourning. The disturbance of self-regard is absent in mourning; but otherwise the features are the same. Profound mourning, the reaction to the loss of someone who is loved, contains the same painful frame of mind, the same loss of interest in the outside world – in so far as it does not recall him – the same loss of capacity to adopt any new object of love (which would mean replacing him) and the same turning away from any activity that is not connected with thoughts of him. It is easy to see that this inhibition and circumscription of the ego is the expression of an exclusive devotion to mourning which leaves nothing over for other purposes or other interests.

An important claim of Freud's is that if one listens patiently to a patient's many and various self-accusations, one cannot in the end avoid the impression that often the most violent of them are hardly applicable at all to the patient himself, but that with insignificant modifications they fit someone else. The patient, Freud claimed, is accusing not himself but some lost loved person or object. Freud assumed that all love is ambivalent in depressives – or melancholics as he called them – and that hostility towards the loved object is turned inwards. Thus a patient who is depressed is mourning for someone who is consciously or unconsciously believed to be lost. The loss is a real one in the case of bereavement. But more often the patient is angry with the loved one and wishes that person to be dead, kills the person in fantasy, and then mourns the loss.

Later analysts, such as Abraham in the 1920s, proposed a theory in which lack of oral gratification was central, together with infinite disappointment in relation to the mother. It was as one commentator put it, as if he thought of melancholia as some sort of mental indigestion. Sandor Rado developed the theory, emphasising the depressive's intensely strong craving for narcissistic gratification; he compared a depressive to a young child who is overdependent on his parents' approval. It is for this reason that the depressive cannot tolerate disappointments. The psychoanalyst Bibring emphasised the importance of the 'ego' in the pathology of depression; he saw depression as the result of a conflict within the ego that leads to feelings of helplessness and powerlessness because of the discrepancy between one's goals and the ability to achieve them. These goals include the wish to be worthy and loved, strong and superior, and good and loving. He rejected the suggestion that depression was due to oral fixation at an early age, although he did suggest that frequent frustration of oral needs may initially cause anger and then be replaced by feelings of helplessness. But I agree with those who have pointed out that terms like 'ego' are rather arbitrary, poorly defined abstractions, not structures.

The analyst Jacobson, for her part, emphasised that loss of self-esteem lay at the core of depression, arguing that during development the child's self-image is constantly changing and it is possible within a loving family to develop an optimal level of self-esteem. Melanie Klein claimed that children experienced a depressive position at the time of weaning and this resulted in rage and sadness at the apparent loss of the mother. She thought that depression in later life was related to the inability to deal with this early loss. René Spitz, by contrast, denied the existence of the depressive position in normal development but emphasised the severe distress that occurred when there was an abrupt separation from the mother. Another psychoanalytic theory emphasises the role of fears about castration, even claiming that this could account for the greater prevalence of depression in females: if a woman envies men their penises and feels that she has been 'castrated', this sense of loss could have a depressive effect.

Interesting and imaginative as these ideas may be they are often difficult to understand and, more importantly, impossible to validate or disprove, both because the concepts are so poorly defined and also because they are reconstructed from the analysts' interpretations. There is no scientific study that provides any basis for their validity. Nevertheless, they have put great emphasis on the importance of loss and the role of early experience.

John Bowlby's ideas on attachment and the effects of loss grew out of psychoanalytic theory, but were also influenced by his observations of children and studies on animal behaviour. His early work emphasised that children's experiences of interpersonal relationships are crucial for their psychological development. Human beings instinctively forge strong bonds of affection. This is particularly clear in young children who suffer considerable distress – sadness and anxiety – when separated from their parents. Rather like grieving, an initial period of protest is followed by searching, and the failure to find the loved object can lead to excessive, even malignant, sadness.

While the behavioural and emotional tendencies associated with attachment are essentially instinctual, they are modified by learning, particularly during childhood. The infantile urge to attach ensures children's safety until they are old enough to survive without parental care. As a child grows older the strong attachment to its parents diminishes, and new attachments are made, usually to a person of the opposite sex. This later attachment is also adaptive as it helps maintain the mutual support needed to raise children. Sadness at the loss of the person to whom one is attached is a necessary consequence of the attachment instinct; the avoidance of sadness helps to maintain the attachment.

Bowlby maintained that attachment theory was fundamentally different from psychoanalytic theory as it rejected the idea that an individual passes through a series of stages in any one of which the child may become fixated and to which the adult may later regress. Bowlby also rejected the idea that emotional bonds were derived from drives related to food and sex, and the theories of Melanie Klein, since they were based on unconscious fantasies and intrapsychic conflict, the evidence for which was effectively non-existent. Bowlby considered the working mental models that were developed through attachment to be central components of personality. These working models of how others will respond, and how one perceives oneself, are enduring; it is not clear, however, to what extent they might be modified by later experiences.

Attachment theory and the less conservative and abstruse modern psychoanalytic theories have features in common. Both view infant–mother attachment as an autonomous motivational system. Both assume that in order to experience oneself as a separate person comfortable in exploring the world, one must have experienced adequate early caring and have been able to 'internalise' this experience. Central to attachment theory are three ideas which can be traced back to Freud: a belief in the powerful influence of the parent's behaviour on the child's personality and social development; a belief that this influence is established early on, and that the relationship with the parents or care-givers provides the child with a mental model that governs further interactions; and a belief that the experiences which shape the mental model and then activate it include anxiety-provoking interactions caused by the loss of love or a loved person.

Animal studies were an important influence on, and support for, Bowlby's attachment theory. A six-month-old monkey's response to separation from its mother is remarkably similar to that of a child. There is agitation and exploration accompanied by screams and cries. Play with friends ceases and objects in the room are ignored. At the time of going to sleep there is an increase in crying and agitation. Over the next few days there is withdrawal and the animal becomes lethargic and unresponsive. This depressed pattern persists for some time. Transfer to an unfamiliar environment at the time of separation makes things worse.

The general developmental picture from studies of Old World monkeys and apes is similar. Virtually all the infants spend their initial days or months in almost continual contact with their mother and usually cling to their mother's chest for much of the day. They consistently display most of the instinctual responses found in human babies – sucking, clinging, crying and following.

Smiling is seen only in chimpanzee infants. After this initial phase all the infants use their mother as a secure base from which to explore their environment.

Infant rhesus monkeys spend almost all of their initial weeks of life in physical contact or at least within arm's length of their mother and thus form an enduring attachment bond with her. It is only in their second month that they begin actively to explore their environment but almost always they use their mother as a secure base. At about six months their play with other monkeys of the same age becomes the predominant activity. Males leave the group while females remain together for the rest of their lives and the attachment to the mother persists into adulthood.

Monkeys that have been separated from their mothers at birth and hand-reared in a nursery for their first month and then placed in small social groups with whom they grow up – peer-reared monkeys – develop strong attachment bonds with each other. However, these attachments are not as strong as in maternal development, and while otherwise normal these monkeys show less exploratory behaviour and tend to be shy and anxious.

About 20 per cent of the monkeys in a typical rhesus colony are from an early age much more sensitive to novel stimuli and are highly reactive, often showing fear and anxiety with minimal provocation. They are also less exploratory. If these monkeys experience poor maternal care they are more likely to suffer depressive symptoms, becoming passive and anxious, and adopting a motionless foetal posture away from the rest of the group. There is reason to believe that the high reactivity is largely genetically determined and these observations thus provide an important example of the interaction between genetic and environmental influences.

Attachment theory has naturally led to a great deal of attention being focused on how securely attached children are to their care-givers. A virtue of attachment theory is that it provides, just as John Bowlby hoped, a basis for experimentation and testing of the associated ideas. One of the experimental set-ups used to study attachment is the Strange Situation. A mother and her one-year-old child are introduced into a playroom which they have not been in before. After twenty minutes, the person carrying out the experiment then asks the mother to leave the room for three minutes, leaving the child with the experimenter. After the mother's return, both mother and experimenter leave the room and then return to the child after three minutes, reuniting mother with child. The whole process is recorded on video tape and the responses of child and mother analysed with particular focus on how the child behaves when separated from its mother.

There is in the child a complementary relationship between attachment and exploratory behaviour, the latter taken to reflect a secure attachment. If children experience their care-givers as available and responsive and so feel secure, they will move further and further away from the care-giver as they explore their environment and even establish contact with others who are present. This relationship has been systematically investigated in the strange situation: it is claimed that the healthiest attachment is shown by a secure child who has a high level of animation, organisation and investment in play when the parent is in the room. Insecure children are much more restricted in their activities. Abnormal behaviour in the strange situation is claimed to be a predictor of later psychological difficulties but evidence based on such a short period of observation is weak.

Children respond differently, but four main patterns of response have been identified. In what is called secure attachment the child is usually distressed when separated from the mother, but greets her at the reunion, is comforted quite quickly and then returns to contented play and exploration. In the insecure-avoidant response the child shows little sign of distress when the mother leaves and ignores her when she returns. When she is present play is restricted and the child keeps watching her. By contrast, in the insecure-ambivalent response the child is very distressed at the separation and is hard to comfort at the reunion. Typically the child seeks contact but then resists, turning away, and alternates between anger and clinging, with little exploratory play. The fourth type of response is regarded as abnormal and is called insecure-disorganised. The child shows a diverse range of confused movements which include becoming almost frozen when the mother returns. The Strange Situation has been used by many investigators in different cultures and the same patterns have been observed though the frequency varies. For example, insecure-ambivalent is more common in Israel and Japan than the USA and Europe, where insecure-avoidant is more common.

Observation of mothers with their children has indicated that the mothers of secure one-year-olds have been responsive and in tune to the child's needs, whereas mothers of insecure-avoidant children are much less responsive, and insecure-ambivalence reflects a mother's inconsistent response. There is some evidence that poor childhood attachment experiences lead to vulnerability to depression in adult life. In insecure-ambivalent attachment sometimes the child's needs are attended to and at other times not; sometimes the response is rapid, other times slow; on occasion the response is over-indulgent; the parent is intrusive and unpredictable. Children brought up in this way are claimed to have been unable to find an effective way of signalling their needs or feelings to others. They remain dependent on their care-giver and fail to develop a proper sense of their own abilities and so lack self-esteem. In later life they become preoccupied with the belief that relationships are not effective. Such individuals may be vulnerable to becoming depressed, their depression is characterised by relying heavily on others but feeling lonely, rejected and unloved, though the evidence for this is merely suggestive.

In avoidant-attachment, the parent or care-giver behaves in a predictable manner but in a way that increases the child's discomfort, often by not responding at all. The children thus learn to ignore their own feelings to avoid reactions from their parents that they find unpleasant. They are prone to self-critical depression, in which the individual feels worthless and constantly tries to compensate for that feeling. They invest in achievement rather than relationships. Loss of control worsens their self-esteem. Again, though, the evidence for this is no more than suggestive.

The attachment patterns established in infancy are relatively stable and over 80 per cent of the behaviour of one-year-olds was similar to their behaviour five years later. Secure infants at the age of six were more socially confident, were better at concentrating and were more positive than those who had not had a secure attachment. Moreover the patterns appear to carry over to the next generation. The attachment status of a mother, discovered through interviews, gives a good indication as to how her child will respond. Parents who give a coherent and detailed account of a secure childhood generally bring up secure children. Mothers who are described as dismissing-detached give brief, incomplete and often idealised accounts. Such mothers tended to have ambivalent-attached infants. There is also evidence that depressed mothers have difficulty generating secure attachment in the child.

While there is a natural tendency to emphasise the role of the mother in these studies, it is most unwise to neglect intrinsic biologically based differences in infants, whether rats, monkeys or humans. Some infant monkeys, as we have seen, are born abnormally fearful and even freeze quite still with very little provocation. In humans, too, there appear to be innate temperaments. Children were examined from 16 weeks of age at intervals until seven years old. About a fifth of the 16-week-old infants were highly reactive to tests such as the smell of alcohol – it made them fretful and agitated. About a third showed little or no distress. With time, some of those who were initially highly reactive began to respond normally but none became a bold, fearless child. By age seven, one third of the highly reactive group had developed extreme fears, more than threefold the rate in the other groups. So a vulnerability may be present from birth, possibly genetically, possibly due to influences during embryonic development.

At this stage one should avoid the temptation to put too much emphasis on early childhood experiences as a major causal factor for depression, for attractive and seductive as such an idea is, the evidence to support it simply does not exist, other than in severe cases like child abuse, severe family discord and loss of a parent followed by inadequate parenting or care.

Bereavement may provide a key to understanding depression, as it generates intense sadness owing to the loss of a key attachment. As Freud made so clear, there are striking similarities between depression and bereavement; key differences are that the sadness of mourning is not associated with loss of self-esteem, and there can be a feeling of anger with whoever is held to be responsible for the death: the responsible agent may be someone outside the group, the dead person, or even the mourner. But the significance of the similarities is such that unless we can understand mourning we are most unlikely to understand depression.

Bereavement results in many of the feelings associated with depression. It is commonly followed by a period of shock and numbness. Grief can be intense and there is a feeling of emptiness not unlike depression. Episodes of depressed mood and anxiety are common. Loss or disappearance of someone to whom one is attached, whether parent, lover or friend, provokes a powerful urge to find that person. The failure to do so results in distress and pining. In the case of death, even though the person is gone for ever, there is still an urge to find that person. The fruitless search results in the pain of grief which, with depression, is probably the most powerful psychological pain in human experience.

Sanders has proposed an integrative theory of bereavement incorporating the ideas of other workers in this field. She identifies five phases of bereavement. In the first there is shock, which can protect the griever from being overcome by the pain of separation. In the second phase the shock wears off and the reality of death becomes clear, leading to severe stress. In the third phase, there is a tendency to withdrawal and the start of the fourth phase, healing. In the final phase there can be renewal and resolution of the grief.

Chronic mourning can result when grieving continues over a period of a year or more. It can lead to chronic despair almost indistinguishable from major depression. Bowlby suggested that childhood experiences of the type that can lead to chronic mourning can also lead to depression. Among these experiences is that of never having had a stable and secure relationship with parents or having lost a parent during childhood. Parkes presents evidence that the patterns of attachment formed in childhood affect relationships in later life and these in turn influence the responses to bereavement. For example, children with anxious attachment become very anxious and insecure after bereavement.

The rituals in different cultures for dealing with bereavement are very varied. In spite of the probably similar biological basis of responses to bereavement, cultural influences can have a profound effect on the course of grief and how it is expressed. The expression of grief in public is an accepted ritual in many cultures and is often seen as therapeutic. Puerto Rican women are expected to express their sorrow with violent emotions; some Asian groups, while wailing in public, are expected to be stoical in private; Greek and Portuguese women are traditionally expected to grieve for the rest of their lives, and ritual suicide on the funeral pyre of a dead husband, though illegal, is still practised in India. By contrast, in Western societies mourning is perceived as a phase to be got through as quickly as is decently possible. It is common for mourning rituals to be prescribed to only last a certain time, for example, one year in traditional Jewish culture. In Japan, mourning rituals encourage a continuing relationship with the dead person, and some sort of small altar – just a photograph will suffice – is made in the home.

One of the terrible features of bereavement is that there is nothing that the bereaved person can do to restore the lost individual; there is loss of control. Laboratory experiments have shown that when an animal has experienced a trauma that it cannot control, its motivation to escape, and so avoid the trauma, wanes. Dogs given an inescapable and moderately painful electric shock usually develop what is known as learned helplessness, rapidly giving up any attempt to escape even when the possibility exists. In Seligman's view when a person learns that an outcome is independent of his or her response, it has a profound emotional effect. If, for example, one learns that relief from an unpleasant stimulus is not dependent on one's own actions, one stops responding, and becomes more passive. Perhaps this is related to one of the stages in mourning and the gradual cessation of 'searching' for the lost person. One can also see this in an infant monkey separated from its mother; it initially searches frantically and makes distress calls, but eventually takes up a passive position.

There are similarities between the symptoms of learned helplessness in animals and depressed patients; for example, passivity, learning difficulties, weight loss. Seligman suggests depression is thus due to the belief that action is futile, and that nothing can be done to restore the losses linked, for example, to bereavement, financial difficulties or chronic illness. This shows how the loss of control over external events that results in sadness can distort the way a person thinks, though the feeling that one has lost control over one's fate and is in a helpless situation may be more imagined than real.

Everyone wants explanations for the events that affect their lives. How individuals account for events that affect their own and other people's lives is a well-studied area of psychology known as attribution theory. For example, it is quite common for people to explain their own behaviour in terms of external events, but the behaviour of others as being due to internal factors, such as their personality. The habitual explanatory style that depressed people bring to bear on events they experience has been linked to learned helplessness. People have, for example, a general belief as to why certain events are rewarding; some believe that the rewards are internal, that is due to their own efforts and abilities, while others believe that they are external, due to other people or chance. This distinction is also true for the way they view uncontrollable distressing events. A further distinction is between causes which are stable or unstable, a distinction that reflects a belief as to whether things can change. For someone who has suffered a rejection in a romantic relationship an internal stable explanation would be 'I'm unattractive', while an internal unstable would be 'sometimes I am boring'. By contrast, an external explanation would be 'romance is difficult for him or her'. A further relevant characteristic mode of explanation is the distinction between global and specific explanations. A global explanation for a student failing an exam would be 'stupidity', whereas a specific one would be 'I am not good at multiple-choice exams'. Global, stable and internal attributional thinking is characteristic of depression.

Seligman suggests that the feeling of loss of control can also arise from receiving positive rewards for which one has made a negligible contribution. It is essential to our well-being that we feel we control our own lives. So perhaps the reason that very successful people get depressed is that they are being rewarded for what they did in the past rather than current achievements. For instance, someone may think that there are more beautiful women who are depressed than one should expect, but the women feel that they are rewarded for their looks alone, and not for something they themselves have achieved. Another example of the importance of controllability is the hypochondriac whose anxiety about their heart is allayed by carrying pills that they believe will work.

While a common view is that depressives have a distorted, negative view of reality, there are some studies which suggest that this might not be quite as it seems. In experiments on students in which they had to assess the amount of control they had over whether a light was on or off, depressed students did much better. Again they were better at predicting academic success – non-depressed students were over-optimistic. Unlike depressives who explain good and bad events with the same explanatory style, non-depressives often see success as internal and stable and failure as external and unstable.

There is a strong link between attachment theory and Aaron Beck's most influential cognitive theory of emotional states, especially depression. It is a theory that can be thought of as dealing with the malignant invasion by emotions, particularly sadness, of cognitive processes, together with the effect of conscious thoughts on the emotions themselves. Cognitive theory says that certain beliefs about ourselves, other people and events develop from an early age. These beliefs are like the models referred to in attachment theory and are encoded at an early age in terms of phrases like 'I am unlovable' and 'I am incompetent'. These beliefs may be generated by the nature of the attachment that develops in infancy as well as later experiences including separations and other losses and traumas.

Beck has proposed the term schema to describe the patterns of thinking that maintain a depressed patient's self-defeating and pain-inducing attitudes even when objective evidence for positive factors is present. The assumptions in a depressed individual are dysfunctional because they are rigid and not necessarily related to reality. For example, if someone believes he or she is incompetent there is a constant striving to be successful at all times, even if the goals set are unrealistic. The so-called cognitive triad has been used to characterise depression and consists of the individual's dysfunctional view of the self, of the world and of the future. In the inner world of the depressive, the self is seen as defective and inadequate; the outside world is seen as presenting insuperable obstacles; and there is the conviction that the depression will continue for ever – that the future is hopeless. An essential feature of the theory is that such negative thoughts are automatic.

When, some 25 years ago, Aaron Beck was developing his ideas about cognitive therapy, the main theories about emotional disorder shared the assumption that the patient is under the control of concealed negative forces over which he or she has no control; for example, that unconscious elements are sealed off by conscious barriers which only psychoanalysis can penetrate. Beck came to the conclusion that this was wrong, as it failed to take into account the person's conscious thoughts. Beck was himself a psychoanalyst and had been practising psychoanalysis for many years before he realised that the patient's conscious thoughts could have an important effect on both emotional states and behaviour. He found that patients had two simultaneous streams of thought, one of which might be hostility towards the analyst and which was expressed, while the other, unexpressed, was self-critical, for example, 'I have said the wrong thing to him . . . I'm bad'. These negative thoughts were rarely noticed by the patient and seemed to happen automatically, almost autonomously.

Beck proposed that the patient is, in principle, able with help to understand and deal with a psychological disorder like depression. In these terms the aim of the therapist is to help the patient unravel distortions in thinking and to learn alternative, more realistic ways to formulate and deal with problems. This again is in marked contrast to psychoanalysis which usually regards the patient's explanations as spurious rationalisations. Cognitive therapy tries to induce the patient to use the same problem-solving techniques that are used in day-to-day life.

Blunting of the capacity to experience pleasure is a key feature of depression. This may be due to the inability to learn to respond to pleasurable stimuli. According to Beck, depressed patients have cognitive distortions about themselves. These include: drawing negative conclusions without any evidence to support them ('I failed once, this means I will never be successful'); focusing on negative details and ignoring positive ones ('I am just a mass of problems; I've nothing going for me'); reaching major conclusions on the basis of a single event ('John says he can't stand me; nobody cares for me anymore'); relating events to oneself that have no obvious connection ('it was all my fault'); and describing events in extreme terms ('I lost my job, I'm a loser').

Emotional reactions to any event are largely governed by the private meaning of that event; and so excessive and inappropriate anxiety can arise because a set of incorrect meanings have been attached to a particular situation such as flying or entering a hospital. This inappropriate cognition can be illustrated with respect to sadness which results from something of value being lost, such as an object, person or self-esteem. But sadness can come from virtual losses; for example, an anticipated loss, like the future separation from a partner, or the disappointment arising from a discrepancy between what is expected and what is actually received. There can even be a hypothetical loss; sadness at the possibility of losing something.

The continuous downward course of so many depressions, though often initiated by a serious life event, can be explained by Beck in terms of a feedback model. As a result of negative thinking, the patient interprets the loss in a negative way. This can lead to physiological changes such as the inability to sleep, and these too are also interpreted in a negative way, thus feeding back and reinforcing the depressed state. A vicious cycle is set up and maintained. The aim of therapy is to break the link and reshape the patient's erroneous beliefs.

Memory plays an important role in depression and there is preferential access to memories of negative events. It is likely that part of the depressive condition reflects mental activities of which we are not conscious. Memory for some aspects of the past can be expressed without any awareness that one is remembering anything. This is known as implicit memory in contrast to explicit, or declarative, memory, to which we have ready access. Direct evidence for the two kinds of memory, implicit and explicit, comes from studies on patients who have a particular and specific form of brain damage. These patients have difficulty remembering recently acquired information, showing a defect in explicit memory. However, these same patients can perform almost normally when asked to carry out tasks which do not require conscious recall, but which can rely on implicit memories.

Emotions and thoughts (cognition) are linked in memory, and so when some event activates a particular mood the events associated with that mood may become accessible. If, for example, one experiences a rejection then, instead of just feeling upset, the response could lead to a depression because the event triggers recall of previous similar events which may be linked to negative thoughts and feelings. Where there is a history of childhood adversity, memories of unpleasant childhood experiences are frequently activated during an episode of depression.

An intriguing aspect of the memory system of depressed patients is the general quality of the memories they recall. Depressed people use an over-general mode for retrieving memories and they are not aware that they do so. The over-general retrieval mode is reflected in the inability of a patient, when questioned by a therapist, to recollect specific events as distinct from general ones. A patient may say that he was happy as a child when his father took him for a walk on the common, but when asked by the therapist to recall a specific occasion, be unable to do so. When given event-words like 'angry' or 'sorry', depressed patients initially give general responses like 'when I've had to do something' and 'when I've had a row'. Control subjects would respond more specifically, for example, recalling being angry with a supervisor on Monday.

Early adverse experiences are linked to a current negative evaluation of one's self. And the over-generalised recall is more pronounced in depressed patients reporting a history of sexual abuse. The reason for this relationship may be that these childhood experiences involve negative attitudes directed towards the person that lead to self-criticism and self-blame. Another explanation may be that during depressive episodes these bad memories pervade, even perhaps invade, consciousness. The over-general type of memory is not found in anxious subjects. It persists in depressed patients even when they have recovered and so seems to be part of the person's long-term style of thinking, and is associated with poor problem-solving ability. Depressed patients perform poorly in short-term memory tests and language comprehension.

In the development of children, a summary style of memory occurs naturally at around three to four years. So three-year-olds can give a general account of what happens at mealtime but are much less able to recall what had happened at dinner the previous day. This is known as 'general event representation', and young children, while capable of recalling specific events, prefer to give general answers. It may be that children who have negative experiences continue to recall memories in general terms in order to avoid the unpleasant memory of the specific events. Adults, when recalling childhood memories, find single events more unpleasant than general memories. It is possible that if the memories of specific events in childhood are too painful then general recall becomes the normal mode for all memories.

It may be possible to develop a theory of depression based on the influence of loss and adverse childhood experiences, triggering in those with a genetic vulnerability severe sadness that leads to long-term cognitive and emotional changes. The sadness could become malignant by 'invading' cognitive processes and distorting them in the manner so characteristic of depression; the sadness prevents any pleasurable responses. If this were so, a particular feature that requires explanation is why depressives are so consistently negative. How could malignant sadness distort their thinking to such a great extent? One possibility that I would like to propose is that these negative thoughts are a type of rationalisation of the patient's emotional sadness. That emotion drives and distorts the person's thinking because the individual adjusts their thinking to fit with the emotion they feel. An analogy may be drawn with the rare and bizarre Capgras delusion. Patients with this illness believe that some of the people they know, sometimes even their loved ones, are 'doubles', not the real person but an alien or robot copy. They assert that there are small but definite differences between the real person and the substitute in order to justify their belief. This example shows the capacity of the brain to function abnormally and distort cognitive functions in ways that seem real and rational to the patient. It would be satisfying if one could identify a biological basis of depression – to establish that there were indeed changes in the brain associated with the emotion of sadness and the stresses caused by loss, and which interacted not only with each other but with the cognitive processes that produce negative thinking. So, are there processes and structures in the brain that could account for this? As proposed by Beck, there must be powerful interactions between cognition and the biological basis of emotion.

CHAPTER NINE

Biological Explanations and the Brain

The brain, to put it mildly, is very complex and contains an astonishing number of nerve cells and connections. Even so, it is possible to link some psychological processes to processes within the brain. To understand depression it would be both satisfying and helpful, for example, to find evidence that early experiences in relation to attachment and loss could establish long-term effects in the brain; is there biological evidence for implicit memories affecting emotions like sadness? We want and need to know how emotions and thinking can affect the chemistry of the brain and how the chemistry of the brain affects our thinking and feeling. Does stress, for example, have long-term effects and, if so, how are these produced? And how could sadness be malignant?

Schizophrenia can provide a good model for the study of mental illness. In the USA it affects about one in a hundred people. It used to be thought to be caused by psychological stress but the generally accepted view now is that it is a disorder of the brain, that is, it has an almost entirely biological origin with a strong genetic component. Studies on the structure of the brains of schizophrenic patients have revealed consistent abnormalities. Are there abnormalities in the brains of depressed patients?

Neurons, the nerve cells in the brain, have long extensions which can receive inputs from as many as a thousand other neurons, and usually just one long extension which transmits its signal. A sense of the numbers and complexity of the brain can be gained by imagining it being increased in size so that the central region of the neuron, which contains the nucleus, a small bag containing the DNA, is about the size of a human body. There are, it is estimated, about a million million neurons in the brain, more than a hundred times the number of people on earth. If the brain is enlarged so that the cells are human in size, it would cover a big block about ten kilometres in all directions, so covering much of Manhattan and extending way up into the sky. What is more, each of these billions of neuron-people would communicate on average with about another 100 to 1,000 other neuron-people. That grotesque image might make it clear why the brain is complex and difficult to analyse. To make it even more complicated, if that were possible, there are even more cells of another type, the glia. Glial cells nourish, interact with and insulate the neurons.

One has to think of the brain in terms of these billions of neurons interacting with one another. The circuits of such interactions determine how we think and behave because the connections between the neurons determine how the brain functions. Many of these connections are set up during embryonic development and are controlled by the genes, and are then modified by experience. Not all neuronal connections are made before birth and the numbers continue to increase until about the age of 11. Learning involves the changing of the connections and their strength. Yet among all these billions of neurons it is possible to find single cells that become more active when we see a particular face. The relationship of the activity of single neurons to the larger community of neurons presents profound problems, and neuroscientists, in spite of some remarkable advances, cannot pretend that they understand many aspects of brain function yet.

Discussing the biological basis of depression and brain function requires the introduction of terms and concepts much less familiar than others in this book. No one has difficulty in thinking about the role of life events or early childhood experiences, but chemicals like catecholamines, cell structures like autoreceptors, brain structures like the hypothalamus and amygdala and hormonal systems like the HPA may initially present some problems.

There are two main subdivisions to our nervous system, the voluntary and the involuntary, or autonomic, systems. The voluntary is the one we consciously use to move our muscles, speak and so on. Our autonomic nervous system is not under our conscious control and it controls the movement of many internal organs like the heart, blood vessels and gut. It also controls the secretion of hormones by key glands such as the pituitary at the base of our skull and the adrenals, attached to our kidneys. Much of the biology of depression is associated with the involuntary system.

The brain seems to be built up out of modules; that is, different functions are partly localised in special regions, like the well-known language centre which, if damaged, renders a person unable to speak. The brain thus has a number of discrete regions and structures whose different functions could help us to understand depression. At the base of the brain are two structures that release hormones, the pituitary and the hypothalamus, both of which lie just above the roof of the mouth. On either side of these structures lie the two regions of the amygdala, an organ shaped like a walnut and of similar size, which plays a crucial role in emotion. Close to the amygdala is the hippocampus, which is involved in the making of memories. The cortex is the outer layer of the brain and the prefrontal cortex is expanded in primates and humans and is associated with higher mental functions. The cortex contains all the regions involved in movement of muscles and perception of external stimuli. The regions of the cortex which lie on the sides, under our faces, are particularly involved in cognitive processes and social interactions. Beneath the cortex and lying above the hypothalamus is the anterior cingulate which is involved in conditioning and receives input from the amygdala. In the brain stem, near where the brain joins the spinal cord, are two regions that send signals to other regions of the brain; the locus coerulus secretes noradrenaline and the raphe nuclei secretes serotonin.

All these regions have complex connections and interactions with other parts of the brain. Nevertheless, a very simplified view of depression would be that it is the result of thinking in the cortex continually telling the rest of the brain that something very stressful is occurring, and the activity in the cortex is maintained by input from regions like the amygdala. In very severe cases of depression for which other treatments have failed, the connections between the cortex and the rest of the brain may have to be cut to relieve the condition.

There is little doubt that changes in brain chemistry are linked to depression, particularly the levels of neurotransmitters and hormones. Neurotransmitters are chemical signals, but unlike hormones they do not, in general, circulate throughout the body, instead they act over very short distances. They provide the signals by which the neurons communicate with each other. Neurons have long wire-like extensions along which an electrical signal can be sent when the neuron is stimulated to fire. When the electrical signal reaches the end of the extension where it contacts another neuron, the electrical signal is not passed directly onto the neuron with which it is in contact, but at the junction there is a specialised region known as the synapse. When the electrical impulse reaches the synapse it causes the release from the neuron of a neurotransmitter, a small molecule, which diffuses a very short distance to an adjacent neuron where it binds with receptors on the surface and so can stimulate that neuron to fire or, depending on the neurotransmitter, inhibit it from firing. Two neurotransmitters repeatedly implicated in the cause of depression are serotonin and noradrenaline.

The state of the brain is determined by which neurons are active. When we see a light it stimulates neurons in the retina to fire and these in turn set up complex patterns of activity of other neurons in the brain. When we move our hand it is because neurons in the brain fire, leading to neurons that stimulate arm muscles to contract in the correct way. Thus, the state of the brain is determined both by the electrical activity in the neurons and the chemical signals they release. If all receptors for the neurotransmitters were blocked or neurotransmission eliminated, the brain would cease to function.

A widely held view is that depression involves a deficiency of neurotransmitters like noradrenaline (epinephrine) and serotonin in the brain. After the neurotransmitter has been released and some of it binds to the receptor on the adjacent neuron, it is broken down, for otherwise it would continue to stimulate the neuron. In some cases the free neurotransmitter, for not all of it is either bound or destroyed, is taken up by the neuron that released it and used again. If this uptake is blocked there will be more neurotransmitter in the synapse and thus the neuron will be more strongly stimulated. It is the blocking of the uptake of neurotransmitters like serotonin that is thought to be the basis of the action of antidepressant drugs. We will return to this after a consideration of brain functions which relate to emotion and experience.

The nerve cells in the brain are exposed to a wide variety of chemical signals and one class of molecules that is thought to play a major role in depression is hormones, which also circulate throughout the body via the blood. Hormones can cause significant changes in bodily and brain function, and this is particularly clear in relation to stress. Stress is a general term that refers to any excessive demand, either physical or psychological, that is placed on an individual, and ranges from fear to life events. Events that cause a loss can result in the stress associated with sadness. Coping with stress requires evaluating and adapting to a situation. The ability to deal with both physical and psychological stress varies considerably. Some people treat stress as a challenge which they almost enjoy overcoming, while others feel overwhelmed by stress and unable to cope. The relevance of stress to depression is that stressful life events are associated with the onset of depressive episodes. Since the brain is obviously involved in responding to stress and it is well-known that stress causes hormonal changes, it is of great importance to know how these hormonal changes influence brain function and are involved in the ability to deal with stress.

In an emergency reaction, like that in response to a frightening stimulus, the body is mobilised for fight or flight by the autonomic nervous system. The physiological response to stress is an increased heart rate and raised blood pressure due to the release of adrenaline and noradrenaline from the adrenal gland. The hormone cortisol is also released from the adrenal gland and is a integral part of the stress response. Adrenaline and noradrenaline belong to the class of signalling molecules known as catecholamines and they are both hormones and neurotransmitters. Adrenaline helps deal with crises by increasing the flow of blood to the muscles. Another, but slower, response to the fear stimulus is the release of steroid hormones known as glucocorticoids which include hydrocortisone and cortisol. Whereas adrenaline acts in seconds, glucocorticoids act in minutes or hours. It is not really clear just what the role of raised cortisol is in helping to deal with stress but it does assist the body in mobilising energy reserves and it acts as an anti-inflammatory agent. (It is the latter action for which cortisone, also a glucocorticoid, is used as a drug to relieve pain and inflammation in many conditions such as arthritis.)

This release of cortisol into the bloodstream is due to ACTH (adrenocorticotrophic hormone) which has been released from the pituitary, acting on the adrenal gland. The sequence is as follows. When something stressful happens, processes in the brain cause the hypothalamus to secrete CRF (corticotrophin releasing factor) into the pituitary circulatory system which lies close to it. Within about 20 seconds CRF triggers the pituitary to release the hormone ACTH which can then cause the adrenals to release adrenaline and cortisol. This set of interactions between the hypothalamus, pituitary and adrenals is often referred to as the HPA system.

Normally cortisol levels are high in the morning, presumably to prepare for the day's activities, and low at night when the body rests. But abnormally high concentrations of cortisol can cause depression and this is of great significance as it shows that depression with all its psychological features can have a purely biological origin. There are receptors for cortisol on neurons in the brain and high concentrations of cortisol can induce significant changes in mood, and can cause depression. The evidence for this comes from patients with Cushing's syndrome in which there is a very high concentration of cortisol and the individual becomes obese, has high blood pressure, is easily bruised and has a variety of disorders of the blood vessels and bones. More than half of such patients become depressed. The high cortisol level can be due to a tumour of tissues that produce ACTH or cortisol itself. Patients whose illness requires them to take drugs that raise cortisol over long periods also have a high probability of becoming depressed. It is very significant that about half of patients with severe depression have raised cortisol levels.

It was initially thought that raised cortisol could be used to diagnose depression using the dexamethasone test. This tests for whether control of cortisol levels is normal and makes use of a synthetic substance that resembles cortisol, dexamethasone. When dexamethasone is injected, the brain responds as if the level of cortisol has been increased and so reduces cortisol secretion. Thus, in normal people, the day after a dexamethasone injection cortisol levels are down. However, about 50 per cent of depressed patients do not show this suppression. It was hoped that this lack of suppression could provide a biochemical test for depression, but unfortunately no suppression also occurs in people suffering many other conditions, such as those who have recently lost weight, are physically ill or simply stressed by being in hospital, so its value is dubious.

Why is the concentration of cortisol high in many depressed patients? It could be due to increased stress, and in depressed patients there is an increase in the number of neurons in the hypothalamus that secrete CRF, causing the pituitary to release ACTH which causes the adrenals to secrete cortisol. The amount of this releasing factor is increased by stress, and blocking its action reduces the fear or anxiety associated with stress. CRF has a higher than normal concentration in the spinal fluid of depressives and could well be a major factor in maintaining the depressed state. Studies have found that in subordinate animals that have apparently given up competing it is the glucocorticoids system that is particularly active.

How does cortisol affect the brain? There is some evidence that high concentrations can damage neurons in the hippocampus and this could affect aspects of memory. Glucocorticoids, like cortisol, may affect mood states by their action on the synthesis of serotonin or the serotonin receptors. Serotonin function is reduced in the presence of high cortisol. The effect of stress on the brain is undoubtedly more complex and stress can also result in an increase in the activity of certain genes in the brain. Another hormone known as DHEA (dehydroepiandrosterone) counteracts some of the effects of cortisol. Its concentration is decreased by stress and is often lower in depressed individuals.

The measurement of hormone levels has been made much easier by the use of a technique that only requires a small sample of saliva instead of blood. This has been used in a study of depression in eight- to 16-year-olds in which low levels of DHEA and high evening levels of cortisol were found in about half of depressed subjects, suggesting again a role for these hormones in depression. Of particular significance is the finding that the relative levels of cortisol and DHEA are good predictors of future depressive episodes and persistent depression. And in further studies of children and adolescents with high risk factors for depression it was again found that the hormone ratios were good predictors of who would have a depressive episode following a later stressful life event.

These are among the major changes in response to a stressful stimulus, but there are many others. The pituitary gland secretes a variety of hormones into the circulation which can stimulate not only the release of adrenaline and cortisol from the adrenal, but also thyroid hormone from the thyroid gland. Which of these stimulating hormones is released from the pituitary depends on signals to the pituitary from the hypothalamus. Prolonged stress can, for example, lead to reduction of activity of the thyroid gland and general shut-down of those functions necessary for growth and reproduction. Low thyroid function is quite common in depressed patients. A test for this, like that for cortisol, involves injecting an artificial version of a hormone that stimulates the release, via action on the pituitary, of thyroid hormone. In about one third of depressed patients the expected rise does not occur.

Stress can also affect the immune system. Major depression is accompanied by signs of suppression of the normal activity of the immune system. This is unsurprising, as a range of psychological stresses are known to suppress immune function. Such stresses range from taking examinations to bereavement. It is likely that reduced immune function is related to the increased activity of the HPA axis, leading to increased levels of the hormone cortisol. However, the relationship between stress and immune function is complex and while some parts of the system may be suppressed others may be enhanced. Moreover, excess production of some of the chemical signals – cytokines – that are involved in controlling the immune response may contribute to maintaining a depressed state or even making it worse. And antidepressants themselves can suppress the immune system.

Cytokines comprise a heterogeneous group of messenger molecules that are produced by immunocompetent cells, such as lymphocytes and macrophages, in order to regulate immune responses. The central action of cytokines may also account for the HPA axis hyperactivity that is frequently observed in depressive disorders. Cytokines may also reduce serotonin levels. Interferon-alpha is a pro-inflammatory cytokine that is widely used for the treatment of a number of disorders including viral infections and skin malignancies. Unfortunately, it frequently induces depression and patients must be treated at the same time with an antidepressant. Although the central effects of cytokines appear to be able to account for some of the symptoms occurring in depression, it remains to be established whether cytokines play a causal role in depressive illness or represent immunological side effects of this disease. Antidepressants may alleviate depressive symptoms by exerting inhibitory effects on immune activation in individuals suffering from major depression.

Somatic symptoms, that is physical symptoms probably caused by the depression, are often the origin of the patient's intention to consult with a physician. Half of depressed patients suffer from pain of some sort. Somatization is thus the tendency to express emotional states, like depression, in physical forms. There have been many attempts to define and measure somatization, characterise patients at risk, and identify possible causal environmental events including cultural factors. But somatization might be nothing else than the outward manifestation of sensitisation of the brain to the cytokine system that is normally activated in response to activation of the innate immune system, and causes the subjective and physiological components of sickness. But this is a complex area still awaiting clarification.

It is not only the stress hormones that may be important in depression but those mainly related to sex. The hormones associated with sex are the steroid hormones, oestrogens and androgens; oestrogens being higher in females and androgens like testosterone being higher in males. The difference between the brains of males and females is due to the influence of these hormones acting on the embryo during development. In fact all secondary sexual characteristics, like the development of breasts in females and a penis in males, are due to these hormones. If new-born male rats are deprived of male hormone they will still look like males but will behave like females. If a male child lacks the receptor for male hormone it will have all the outward appearance of a female and the self-perception of being female. It is thus clear that the sex hormones can have profound effects on the development of the brain and on behaviour.

These hormones have a potent effect on mood and mental state. The depressive feelings experienced by some women when they menstruate and after childbirth could be due to the rapid and large reduction in the concentration of oestrogen-related hormones, in particular, oestradiol. It is possible that both oestrogen and testosterone exert their effects by their action on the serotonin neurotransmitter system. But if oestrogen has an antidepressant effect, it raises the question of why women have a higher incidence of depression than men, for they have higher concentrations of oestrogen, and the concentration of testosterone in men is up to 1,000 times the concentration of oestradiol in women. The answer could be that testosterone is converted to oestradiol in the male brain, and so the male brain will actually be exposed to higher concentrations of oestrogen.

Another hormone released from the pituitary is growth hormone. As its name implies, it is an essential hormone for normal growth. It is also present in adults and there are reports of its concentration being abnormal in depressives. Of particular interest is that in breast-feeding mothers it may induce a feeling of well-being, even euphoria. This is particularly significant as the placenta secretes large quantities of growth hormone into the mother's circulation during pregnancy. Thus, after birth the concentration will fall, and this just might provide a biological explanation of postnatal depression.

What evidence is there that early experiences, such as stress, can have long-term effects on the brain and influence brain chemistry? This could provide the link, for example, between attachment, loss and later behaviour. If rats are a good model then the connection exists, without doubt. Human handling of new-born rats produces positive behaviour patterns in the rats' adult life as well as clear changes in their brain chemistry and hormonal levels. Handling simply involves removing the pups from the mother for 15 minutes every day for 21 days. One striking change observed in such tests occurs in hormones like cortisol, whose concentrations are decreased. Handling of the rats after infancy does not result in such changes. Moreover, the rats that have been handled age better and show fewer degenerative changes in the brain; their capacity to learn is also better.

But why should human handling of the rat have such beneficial effects? At first sight it seems odd. It turns out that when the infant rat that has been handled is returned to the mother, the mother doubles her rate of licking and grooming the pup. Handling increases the ultrasonic calls of the pup which results in increased maternal care. It is this maternal response that is crucial. Good evidence for this comes from looking at lots of rats and their pups without disturbing them. Some mothers, about one third, naturally lick their offspring at a high rate, one similar to that induced by handling. The pups of these high-lickers are followed into adulthood and compared with rats whose mothers had licked them much less. The rats of high-lick mothers have characteristics similar to the rats which have been handled in infancy. For example, they are less 'anxious' when placed in an open new environment and so explore it more. They also have changes in their brain biochemistry similar to those that had been handled. The changes in the brain of the low-licked rats involve changes in the number of receptors for neurotransmitters and result in stimulation of the amygdala. Activation of the amygdala is related to anxiety and fear and so this could account for their anxious behaviour. In another study, repeated maternal separation of infant rats made the adults less willing to explore their environment and less able to learn tasks that offered them a reward, a feature common in depressed individuals. These behavioural changes are associated with gross changes in the brain which affect the function of both serotonin and noradrenaline.

It is hard not to be tempted to transfer these findings to the human situation. Does increased stimulation and contact have the same effect on human infants as licking does on rats? Premature babies given additional cuddling and tactile stimulation grow better and mature intellectually more rapidly than those receiving routine treatment. Clearly early stimulation is important for development, and children deprived of a stimulating environment have brains as much as one quarter smaller than normal. It is thus with a feeling of some shock that one recalls the advice of the behaviourist psychologists of the 1940s who advised parents never to kiss their babies or take them on their laps as it might make them too dependent and encourage bad habits.

There may be a mechanism by which acute stressful events during childhood could result in such long-term effects and involves activation of specific genes. The particular chemical implicated is another neurotransmitter, acetylcholine. Acetylcholine is necessary for normal thinking and if its synthesis is inhibited then both learning and memory may be disrupted. There are proteins, known as enzymes, that are involved in the synthesis and breakdown of acetylcholine and the balance of their activity can control the amount of acetylcholine present. Mice that have been stressed by being made to swim for two four-minute sessions initially have an increase in acetylcholine concentration at nerve endings, but then the concentration falls. The fall is due to an increase in expression of the gene in the nucleus that codes for the protein that degrades acetylcholine and so leads to an increase in this protein. Of particular significance is that the activity of the degrading enzyme was still high after three days. These results are relevant to Gulf War Syndrome since they show how stress also involves changes in gene expression. Acute stress, like experiences in a war, can lead to post-traumatic stress disorder, which manifests itself some time after the traumatic event as a series of symptoms that include depression. The demonstration that a short traumatic experience can have long-term effects on the level of a key neurotransmitter makes it difficult to dismiss post-traumatic stress disorder as purely psychological.

A very helpful account of our emotional behaviour and brain function is given by Le Doux in his book *The Emotional Brain*. An absolutely key feature of his analysis is that many emotional states arise unconsciously. This is dramatically illustrated by a patient whose left brain had been isolated by surgery from its right hand side. He could only speak through his left hemisphere, but when an emotional stimulus was presented to the right hemisphere, although he could not say what it was, he could say if it was good or bad; he knew the emotional outcome but not the cause.

It is central to Le Doux's view that we do all sorts of things for reasons of which we are not consciously aware and it is one of the functions of consciousness to tell ourselves consistent stories of why we behave in a particular way. Emotion and cognition – self-aware thinking – are, in Le Doux's view, best thought of as separate but interacting mental functions generated by different but interacting brain systems. Evidence for this comes from the fact that damage to certain brain regions can result in the emotional significance of a stimulus being absent, even though its perception is intact; and likewise that an emotional stimulus can exert an effect without one being conscious of perceiving it.

In general, emotion in relation to brain function has been a relatively neglected area. Because emotions are difficult to analyse many workers have focused on fear, as it is relatively easy to generate in animal models and is easily identified in human subjects. Fear and anxiety are at the core of being human and there has been quite intensive research into the relationship between fear and memory. The storage of a powerful event is called emotional memory and is similar to implicit memory, to which we do not have ready access. Inappropriate fear is tightly linked to a common feature of depression, anxiety. Since anxiety and depression involve the inability of the brain to control fear and sadness, such studies can provide novel insights into the relation between brain activity and depression.

At the level of groups of nerve cells, each emotional unit can be thought of as a set of inputs, an appraisal mechanism, and a set of outputs. The appraisal mechanism ensures that when an animal encounters a natural trigger, like seeing a predator, the emotional response of fear leads to it behaving so as to avoid being eaten. An important concept that is repeatedly used to account for abnormal responses in relation to fear, such as anxiety, is that of conditioning. This derives from Pavlov's experiments at the turn of the century. Pavlov showed that if the sound of a bell occurred when his dogs were fed meat, then the dogs would salivate at the sound of the bell alone. The meat for the dog is known as the unconditioned stimulus and is a natural trigger for some behavioural or emotional response. The bell is the conditioned stimulus, a learned trigger.

Conditioning of rats in relation to fear is a well-studied system. If rats are given a mild shock, they freeze; freezing is the rat's natural response to a danger, like seeing a cat. If they are conditioned with a particular sound they will freeze when they hear the sound. Conditioning can occur very quickly, sometimes even after one pairing of the unconditioned and conditioning stimuli. One can thus think of the conditioning as having set up an emotional memory. Repeated presentation of the conditioned signal without the unconditioned stimulus (the shock) leads to extinction of the conditioned response.

Similar conditioning occurs rapidly in humans, but, as with rats, if the stimulus of light or sound is given repeatedly without the shock, the response diminishes. But has the emotional memory really been eliminated or is it rather that the animal has learned to control the fear? The evidence is that the memory persists and can easily be reactivated; it can even occur again, spontaneously. Once established, emotional memory is hard to lose.

Can implicit emotional memories be eliminated? Different processes are involved in explicit and implicit memory. Explicit memory seems to be mediated by the temporal lobe of the brain; by contrast implicit memory involves multiple systems. While explicit memory is often easily lost and unreliable, implicit, unconscious memory is very long lasting. In the brain there are thus multiple memory systems each devoted to a different function. There is a famous case of a patient with brain damage who could not make new memories; she would need to be re-introduced to her doctor on each occasion she met him. When, one day, the doctor put a pin in his hand so that he pricked her quite sharply when they shook hands, she would not shake hands on the following day. But she had no idea why. The doctor's hand, completely unconsciously, had come to represent danger. Thus, while she had lost the operation of explicit memory she still had the ability to form new emotional memories.

A further example is the experience of a car accident. The explicit memory will enable one to recall the incident when hearing a sound, like a horn blaring, which occurred at the time of the accident. At the same time the implicit emotional memory is activated and the fear and anxiety may again be experienced. In our early childhood, emotional memory, like perhaps a sudden separation from a parent, associated with the amygdala, may develop before explicit memory, which is associated with the hippocampus. This could provide an explanation of how early emotional trauma could, unconsciously, affect our later mental life, just as the psychoanalysts and developmental psychologists have long suggested.

The unconscious has had a major influence on much thinking about mental illness and is heavily mined by psychoanalytic and psychodynamic psychotherapists. A difficulty has always been to get at the unconscious in a way that would make biological scientists more convinced by, or even comfortable with, the concept of unconscious thoughts. The concept of implicit memory is essentially a biological basis for the unconscious. Another set of investigations has demonstrated the role of the unconscious in relation to emotion and awareness and the importance of the amygdala. Regions of the brain, like the prefrontal cortex, prevent emotional responses once they are no longer relevant. Thus, it is likely that if this control by the cortex in some way malfunctions it may be very difficult for a person to control their emotional responses; it is as if a brake on the emotions has been removed. This may be highly relevant to understanding anxiety and may also apply to depression, for early events in childhood may have caused conditioning of an anxious or sad emotion, and this may persist latently well into adult life.

The amygdala plays a central role in emotional responses and learning. An emotional stimulus is passed through another brain structure, the sensory thalamus, to the amygdala, and there are inputs from the sensory cortex. Signals from the amygdala then cause typical responses like freezing, increase in blood pressure and activation of the HPA axis and the release of stress hormones. The amygdala is like the hub of a wheel, receiving stimuli and appraising their emotional meaning. Any damage or malfunction in the amygdala could thus lead to emotional disorders. One patient with damage to her amygdala could correctly identify the emotions associated with various facial expressions, except when the faces showed fear.

One of the studies that have been done to study the role of the amygdala and implicit emotional memory makes use of a technique known as visual masking. A frightening face is flashed on a screen before subjects but only for a very short period, a few milliseconds. This is not long enough for the face to be easily or consciously recognised, but in any case a neutral face is immediately flashed for a longer period and is easily recognisable. This second face completely masks the subject's awareness or recall of the first face. The subjects can only report seeing the second face. Yet the activity of the brain as revealed by imaging techniques showed that the amygdala increased in activity when the fearful face was flashed. Thus we can register fear without being conscious of what has caused it. We have an unconscious recording system in the brain.

This experiment has been extended by coupling it with classic Pavlovian conditioning to an unpleasant stimulus. Subjects were repeatedly shown an angry face which was followed by an obnoxious noise. This conditioning made the subjects have a stronger reaction to the angry face. This response was measured both by changes in the resistance of their skin to a small electric current and also to changes in activity in the amygdala. The subjects were then presented with the face in the masking situation; the angry face was presented very briefly, followed by a neutral face. They did not report seeing the angry face but the amygdala increased its activity and there was a change in skin resistance. This confirmed the previous findings but there was a further surprise. When the fearful stimulus was unconscious it was the amygdala on the right side that increased in activity, but when they were aware of the fearful stimulus it was the left side of the amygdala that responded.

The key role played by the amygdala is shown by patients whose amygdala has been damaged. These patients cannot distinguish between faces which are friendly and those which are hostile; it is as if the amygdala cannot give them the appropriate warning. Monkeys, like humans who are abnormally fearful, have more brain activity on the right-hand side. Depressed patients have greater activity in the right prefrontal region of the brain. Those more fortunate of us who are happy-go-lucky show more activity in the left prefrontal cortex.

Anxiety can be described as unresolved fear. Most theories about the cause of anxiety suggest that it is the result of earlier traumatic experiences, particularly in childhood, that have created an emotional memory of fear but with no explicit memory of the event. But why should explicit memory of the event be lost? The explicit memory of a traumatic event may fail to form because of the stress associated with the event. As we have seen, stress causes release of adrenal steroids. When the amygdala detects danger it sends a message to the hypothalamus which in turn stimulates the pituitary gland to release ACTH into the bloodstream. When ACTH reaches the adrenal gland it causes the release of steroid hormones. If too much steroid is released, the hippocampus, which is central to the formation of explicit memory, fails to function properly, and so the memory fails. Stressed rats are bad at learning. Moreover, human survivors of severe stress have a smaller hippocampus and more memory defects. A further consequence of stress is depression and depressed patients often have poor memory.

By contrast, stress enhances amygdala function and so, possibly, implicit memory. Thus, Freud may have been correct in suggesting that aspects of traumatic experiences are stored in memory systems not easily accessible to our conscious thoughts. But this is not to say that this failure is due to repression. It may be that because implicit memories that underlie anxiety are very hard to extinguish, we may have to learn to live with them, and seek to find ways of controlling them. This has important implications for therapy, for such memories may be permanently inaccessible or effectively so.

We can now view emotions as unconscious processes that do not always give rise to conscious experience. The problem is how emotional information becomes represented in what is known as working memory, the general-purpose temporary storage system that we use when we think. It is likely that emotional reactions depend on the activation of the emotional stimulus of the autonomic nervous system via the amygdala, resulting, for example, in the case of fear, in changes in blood pressure and heart rate and the release of hormones of the HPA system. These bodily signals then feed back on the brain and are an essential feature in the establishment of emotional memories. Given the interactions between the amygdala and other brain functions it may not be too difficult to begin to see how anxiety which is related to the fear of loss could be made malignant by the amygdala being inappropriately active. The overactivity of the amygdala could affect many other brain systems including those involved in cognition. No doubt similar processes will be discovered in relation to sadness itself.

The behaviour of the amygdala and the other brain regions involved in emotions and cognitions related to depression are all dependent on neurotransmitters providing communication between neurons. That these chemicals are a major player in depression is very well established, though how they exert their effects is much less well understood; they are relatively small and simple molecules but the effect of changes in their concentration on emotion and cognition can be profound.

Strong evidence for a chemical basis for depression comes from the fact that antidepressant drugs are successful in treating depression and they are thought to act by increasing the concentrations of neurotransmitters like noradrenaline and serotonin in the brain. Certain drugs used to treat other illnesses can induce depression, like cytokines discussed earlier; one of the most significant is reserpine which has been used to reduce blood pressure, and can induce depression in about a third of the patients who take it. My own father was reduced to a weeping shadow of his former confident self when treated with reserpine. Reserpine reduces the concentration of serotonin. It is also striking that mood-altering substances like cocaine and Ecstasy affect both serotonin and noradrenaline. But one must resist the temptation to assume that because changes in concentration of neurotransmitters can to some extent be correlated with depression, they are the cause of the illness. The concept of cause in so complex a system should be treated with care.

Catecholamines comprise adrenaline, noradrenaline and dopamine. The idea that catecholamines are involved in depression was put forward in the mid-1960s. The key idea was that the neurotransmitters noradrenaline and dopamine are deficient in depressed patients. Elation, conversely, would be associated with an excess of such signals. This hypothesis is important not only in its own right but because it was one of the first theories to link a psychiatric disorder with a biochemical change. However, the situation is now much more complicated, as both noradrenaline and serotonin are implicated in the effect of antidepressant drugs, and the two classes of neurons that use these neurotransmitters interact with one another. Neurons that make use of noradrenaline as a neurotransmitter make connections to virtually all areas of the brain. A major region of the brain that secretes noradrenaline is the locus ceruleus located in the hind brain. This brain region is involved in flight and fight responses and regulates levels of arousal, responses of the sympathetic nervous system that include pulse rate and blood pressure and the signalling of danger for the organism. The region itself receives many inputs from neurons that secrete serotonin and dopamine.

Serotonin (also known as five hydroxy tryptamine, 5-HT) is widely distributed throughout the body and is found in high concentrations in the blood vessels, the gut and the brain. In the brain it is mainly located in those regions associated with involuntary activities. The idea that serotonin concentration and function is decreased during depression has been around for over 30 years. Attractive as this hypothesis is to many, especially since almost all antidepressant drugs are thought to act by raising the levels of serotonin in the brain, the evidence is not as conclusive as one would like. One major reason is that it is very difficult to directly measure serotonin levels in the brain. Most measurements are made on its concentration in the blood, which may or may not reflect what is happening in the brain. Only a small percentage of the total serotonin in the body is in fact in the brain. Another reason is that there are at least 14 different receptors for serotonin in different parts of the body, making it difficult to know which are the key ones involved in depression. Nevertheless, there is good evidence that the function of one of these receptors is significantly reduced in the brains of depressed patients.

Neurons that secrete or respond to serotonin are very widely distributed in the brain and this can help account for why alterations in serotonin function can affect so many different behaviours: sleep, learning, movement, food intake and sexual activity. Serotonin-producing cells extend into many of the areas thought to be associated with depression like the amygdala and hypothalamus as well as areas of the cortex. One speculation in relation to depression is that serotonin fibres to the hippocampus have the function of maintaining adaptive behaviours when the organism is faced with threatening situations. Thus, enhancing the response to these nerves could increase resilience and coping behaviour. Serotonin has a significant influence on the release from the pituitary of hormones like CRF, which plays such a key role in regulating the levels of cortisol and related hormones.

The concentration of serotonin in the brain is determined by a number of factors. Serotonin is synthesised from a simple molecule, tryptophan. It is synthesised in the brain and its concentration is determined by the amount of tryptophan that is taken in with food; enzymes, like monoamine oxidase, that break it down; and the activity of the nerve cells. There are special mechanisms in neurons for the uptake of serotonin that is present around the cells. This uptake is due to special transporters of serotonin in the membrane of the neurons and it is on these transporters that antidepressants are thought to act. By blocking the uptake the concentration of serotonin is increased. In addition there are special receptors – autoreceptors – which effectively monitor the amount of serotonin around the cell and act like a thermostat; if the concentration of serotonin goes up the autoreceptor inhibits the synthesis of serotonin, thus trying to keep the level constant. It is this process that has been used to explain why it takes several weeks for antidepressant drugs to have an effect, since it apparently requires a reduction in the number of such autoreceptors, and this is a very slow process compared to the time required to inhibit uptake by the antidepressant. The number of autoreceptors will affect the concentration of serotonin. Also, hormones like oestrogen appear to have an important effect on some of the serotonin receptors and the serotonin transporter. It is clear that the concentration of serotonin can be affected by a variety of factors.

There is evidence that those features of human personality that relate to motivation and emotion may be affected by levels of serotonin in the brain. People with a history of impulsive and violent behaviour, like violent criminals and those who commit suicide by violent means, have been reported to have low serotonin levels. Giving drugs to aggressive psychiatric patients, drugs which increase serotonin function, might reduce hostility and violent outbursts. It is also held that lowered serotonin leads to impaired impulse control and aggression and in depressed patients is associated with suicide. For example, it has been possible to measure the concentration of serotonin on post-mortem brains collected from suicide victims and lower than normal levels are found. But such subjects may have suffered from a variety of psychiatric illnesses and there are many problems associated with changes in the brain tissue following death.

While most studies on serotonin in relation to behaviour have been conducted on psychiatric patients, there are few which have examined the effect of serotonin on apparently normal individuals. One way to try and investigate the possible role of serotonin on mood is by lowering its levels in normal subjects. This can be done by asking them to drink a mixture that reduces the synthesis of serotonin. Some studies on males found that this caused a lowering of mood while others did not. With women there was a consistent lowering of mood, which may reflect their greater vulnerability to depression. The lowering of mood, however, never resulted in anything like severe depression.

In a related approach, serotonin depletion was carried out on patients who had recently recovered from depression following treatment with an antidepressant drug. Many of the patients relapsed into a state with depressive symptoms within a few hours, but recovered when they stopped taking the mixture. However, with depressed patients who were not taking antidepressants, depleting serotonin did not make the depression worse, which suggests that there is no simple relation between serotonin levels and the severity of the depression.

It was thus of interest to see what effect increasing serotonin levels would have on normal humans. Might this lead to a reduction in negative emotions like anxiety and sadness, or increase positive emotions? To find out, an antidepressant drug, Prozac, whose action is thought to raise serotonin levels, was given to a group of normal individuals. The subjects took the drug over a period of four weeks, and then took standard tests measuring 'assertiveness' and 'irritability' as well as 'positive affect' – good feelings – and 'negative affect' or bad feelings. The researchers found that Prozac significantly reduced negative affect – such as fear and anger – while having no influence on positive affect, such as extroversion and optimism. That the effect of Prozac produced these two quite different results suggests that positive and negative emotions might involve different neurochemical systems.

Serotonin has also been found to play a role in the social behaviour of non-human primates. Rhesus monkeys with lower serotonin concentrations show greater aggressive behaviour to other monkeys, receive more wounds and die younger. By contrast, those with high concentrations have a greater number of neighbours, groom others more, and generally seem more sociable. If drugs are given to the monkeys to reduce the level of serotonin, aggression and general anti-social activity increase, while antidepressant drugs that increase serotonin have the opposite effect and increase sociability.

In spite of the enthusiasm and supporting evidence for the role of neurotransmitters like serotonin in depression, one should be cautious at this stage of thinking of low concentrations of serotonin as causing depression. An analogy might be drawn with diabetes, where one would not talk of high glucose being the cause of diabetes – rather, the high concentration can reflect the absence of insulin which enables glucose to enter the cells or, in other types of diabetes, the failure of the mechanism by which it enters cells even in the presence of insulin. The precise role of neurotransmitters in depression is much more complex. This is made particularly clear by the totally unexpected finding that a completely different compound to any of those so far discussed may play a major role in depression. Substance P, so named because it is involved with pain, is a small protein released from many neurons in the brain. But substance P is a neurotransmitter with a much slower action than serotonin. A drug that blocks the receptor for substance P has been found to be a very good antidepressant. It might work by modulating neuronal activity in regions like the amygdala but its relation to serotonin is unknown.

It is possible to examine the living brain in both normal and depressed subjects and see if there are any clear differences in brain function. Changes in local blood flow and breakdown of glucose which are considered to reflect activity in a particular region can be detected by special imaging techniques in the living brain. One technique is known as PET and involves labelling a molecule that is present in the brain, like oxygen or carbon, with a radioactive tracer. When this is injected into the patient its distribution can be followed with special cameras, and computers can work out how its concentration in different regions changes with time. In this way, regions of increased blood flow and glucose usage show which parts of the brain are most active and which are associated with particular mental activities. So it is of great interest to know if there are differences between the activity in the brains of normal and depressed subjects. There is indeed evidence in severe depression that the medial orbital cortex shows increased activity, and that there is a similar increase in activity in healthy subjects in whom anxiety or sadness has been induced experimentally. That it is possible to begin to associate specific regions of the brain with depression is a major advance.

One of the problems in such studies is that the number of subjects studied is usually small and so it is hard to take into account natural variation and perform reliable statistical analysis. Medication can change the pattern of brain activity, and there is also the possibility that brain activity may reflect the particular characteristics of the depressed patients; anxiety and insomnia may give different results from apathy and excess sleep. Attention must be given to changes in the size of different regions which have, for example, been found in depressives who are psychotic or elderly, particularly those with a late age of onset. The results can also be confused by alcoholism. Nevertheless, differences between depressed patients and healthy controls have been identified.

Recent research on the brains of patients who had died and were known to have had major depressive disorder discovered abnormalities in the expression of the fibroblast growth factor system. Fibroblast growth factor protein is a key signalling molecule in embryonic development. Altered expression of the factor was found in frontal cortical regions of brains. Changes in the expression of this factor was reduced by SSRI antidepressants and may thus be partially responsible for the mechanism of action of these drugs. This finding may be related to the theory that growth factor levels positively correlate with a large hippocampal volume, while a small hippocampal volume correlates with susceptibility to stress-induced illness like depression.

The prefrontal cortex occupies about half the volume of the human brain. Within the prefrontal cortex the performance of cognitive tasks or emotional procedures results in increased blood flow in multiple but distinct areas specific to each mental activity. Studies have shown that the dorsolateral regions of the prefrontal cortex have decreased activity in depressives, including manic depressives; moreover, this is reversed by antidepressant medication. Another region with reduced activity in depressed patients is the subgenual region, about 4 centimetres behind the bridge of the nose; and both it and the prefrontal cortex have extensive interconnections with the amygdala and thalamus which, as we have seen, have consistently been implicated as playing an important role in emotional behaviour. Moreover, the size of this region of the cortex is consistently reduced in the left cerebral hemisphere of depressed patients. Still further evidence that this region is involved in depression comes from the observation that it is overactive during periods of mania.

By contrast with those regions of reduced activity in depressives, activity is abnormally increased in the ventral prefrontal cortex as well as in the amygdala and thalamus. Damage to the left anterior prefrontal cortex, for example due to a stroke, is strongly correlated with depression. The amygdala and prefrontal cortical areas excite each other as well as part of the thalamus, and it is plausible that increased activity in these areas is associated with depression. Moreover, surgical treatment for resistant depressions, as well as antidepressant drugs, reduces blood flow in these regions. The amygdala is the only structure in the brain that has been found to consistently show an increased blood flow with depression. It is also of great interest that amygdala activity stimulates corticotrophin release and its activity correlates with cortisol levels. Could the amygdala be the site of origin of malignant sadness?

The significant reduction in size of the subgenual prefrontal cortex in patients with major depression is due to a reduction in the number of cells in that region. However, it is not neurons that are lost, but the supporting cells, the glia, and the discovery of this reduction opens an entirely new way of thinking about the biology of depression. The region of the brain affected is thought to act as a brake on the activity of the amygdala, perhaps by the taking-up of another neurotransmitter, glutamale. Perhaps one genetic influence on vulnerability to depression is related to a reduction in the number of glial cells.

One may thus think about depression in terms of the interactions between activity in specific regions of the brain like the amygdala concerned with emotion, and other brain structures involved in cognition. Life events could activate feelings of sadness, and if the individual is vulnerable this emotional response could be excessive; this in turn could generate negative thoughts and thus further activate feelings of sadness, and so a positive feedback loop is set up, making things worse and worse – malignant sadness.

The weight of evidence is that changes in the concentration of chemicals like serotonin, noradrenaline and stress hormones are associated with depression and that in addition there are changes in the brain that may reflect earlier experiences. While this is excellent progress towards discovering the biological nature of depression, there is still a very long way to go. It is still not known how all these chemicals and brain process interact. Even so, the knowledge about neurotransmitter concentrations in relation to depression, together with the psychological insights that have been obtained, have already been used for the treatment of depression, mainly to break the positive feedback loop.

CHAPTER TEN

Antidepressants and Physical Treatments

Lest one has, with all the discussion about the scientific basis of depression, forgotten just how terrible depression is for the sufferer, and how necessary it is to treat it, let me provide a description by a young woman, Elizabeth Wurtzel:

Some catastrophic situations invite clarity, explode in split moments: You smash your hand through a windowpane and then there is blood and shattered glass stained with red all over the place; you fall out a window and break some bones and scrape some skin. Stitches and casts and bandages and antiseptic solve and salve the wounds. But depression is not a sudden disaster. It is more like a cancer: At first its tumorous mass is not even noticeable to the careful eye, and then one day – wham! – there is a huge, deadly seven-pound lump lodged in your brain or your stomach or your shoulder blade, and this thing that your own body has produced is actually trying to kill you. Depression is a lot like that: Slowly, over the years, the data will accumulate in your heart and mind, a computer program for total negativity will build into your system, making life feel more and more unbearable. But you won't even notice it coming on, thinking that it is somehow normal, something about getting older, about turning eight or turning twelve or turning fifteen, and then one day you realise that your entire life is just awful, not worth living, a horror and a black blot on the white terrain of human existence. One morning you wake up afraid you are going to live . . .

That's the thing I want to make clear about depression: it's got nothing at all to do with life. In the course of life, there is sadness and pain and sorrow, all of which, in their right time and season, are normal – unpleasant, but normal. Depression is in an altogether different zone because it involves a complete absence: absence of affect, absence of feeling, absence of response, absence of interest. The pain you feel in the course of a major clinical depression is an attempt on nature's part (nature, after all, abhors a vacuum) to fill up the empty space. But for all intents and purposes, the deeply depressed are just the walking, waking dead.

Antidepressant drugs are very widely used to treat depression. All drug treatments are thought to depend on their ability to increase the amount of neurotransmitters in the brain, particularly serotonin and noradrenaline. The development of drugs for treating depression, or indeed any medical condition, is very expensive and time-consuming, and a major problem is that the company making the drug will never know the full effectiveness of their drug and, most important, all the side effects, until it has been marketed and used by a large number of patients. Side effects are a feature of almost all antidepressant drugs and these vary enormously. If the side effects are sufficiently serious the drug has to be withdrawn. The story of thalidomide, which was widely prescribed for nausea in pregnant women and caused terrible deformities in the developing embryo, is ever-present in the minds of the pharmaceutical industry, or should be.

The first drug to successfully treat a severe mental illness, schizophrenia, was chlorpromazine, also known as largactil. It had its origin in research in the 1930s into drugs that could help deal with allergies and would act as an antihistamine. When the compounds were being tested on rats it seemed that one of them made the rats lose interest in their environment and become very calm. When tested on humans it seemed to act as a sedative and when tested on patients with schizophrenia was found to be an effective medication. Moreover, the effects were apparent within a few hours of taking the drug. It gave psychiatrists, for the first time, the confidence that other mental illnesses could be treated by drugs.

The development of drugs that can relieve the symptoms of depression has rarely, if ever, been based on rational design – that is, the application of rigorous basic science leading to the synthesis of a particular chemical compound. Rather, progress has come from acute observation by doctors and scientists of humans as well as animal models. Drugs that were being used for other illnesses were often recognised as also relieving depression.

In the 1940s, drugs which had a broadly similar chemical structure to chlorpromazine and which have what is known as a tricyclic chemical structure were being tried out as an antihistamine and also for treating Parkinson's disease. One of these, imipramine, was found to have an antidepressant effect. Imipramine was the first tricyclic to be widely available and has been very widely used. In over 50 studies in which it was compared with a placebo it improved the condition in 65 per cent of patients compared to 30 per cent on a placebo; a similar set of figures comes up again and again with almost every treatment for depression. This was the origin of the generation of a large number of tricyclic antidepressants, some of which are listed in the table below with other types of antidepressants and their side effects. While the side effects are significant, none of the antidepressants is addictive in the same way as, for example, Valium.

The tricyclics, like other antidepressant drugs, are thought to act by binding to receptors on the surface of nerve cells and affecting the local concentration of neurotransmitters. Tricyclics bind to a variety of receptors and block the uptake of not only noradrenaline but also dopamine and serotonin. They also have affinity for other receptors which affect other neurons, particularly for a class of neurons that use acetylcholine as the neurotransmitter, and this accounts for many of their unpleasant and unwanted side effects like constipation and feeling faint. In spite of their side effects, tricyclics are cheap and effective and are thus widely used in countries like Peru, Turkey and India.

At about the same time that the tricyclics were being developed, another class of drugs was being discovered, drugs which inhibit the enzyme monoamine oxidase. The first of these, the drug iproniazid, was used for the treatment of tuberculosis and was found by chance to have a marked antidepressive effect on the patients. Not only did it help with the tuberculosis but it gave the patients renewed energy and a sense of well-being. Hundreds of thousands of depressed patients were treated with it in the first year that it came on the market, both because of the desperate need for an antidepressant drug, and because doctors thought it would be safe as it had been successfully used to treat tuberculosis. But it was withdrawn in 1961 as it was found to cause liver damage. Monoamine oxidase is involved in the inactivation of the neurotransmitter noradrenaline which is a catecholamine. Inhibition of the activity of the enzyme was thought to prevent breakdown of noradrenaline and so increase its concentration. This was one of the main discoveries that supported the idea that depression was due to a lack of noradrenaline in the brain, and that mania was due to an excess of noradrenaline. Monoamine oxidase inhibitors also result in an increase in serotonin. Because of their side effects, they represent only a small fraction of all antidepressant sales. Yet the newer ones with fewer side effects can be valuable in treating atypical depression and patients who do not respond to other antidepressant drugs. They are seldom a first choice treatment.

A newer set of drugs, the Selective Serotonin Re-uptake Inhibitors, known as SSRIs, have been developed since 1987 with more selective actions and fewer side effects. Because they are more selective in the molecules to which they bind, they do not bind to receptors on other classes of neurons like those which use acetylcholine. The most famous of these SSRIs is Prozac.

Large drug companies have revenues of billions of pounds and employ thousands of scientists in their laboratories. The cost of producing a new drug from early research to marketing takes about 10 years and requires an investment of around £200 million. In the late 1960s three scientists at Eli Lilly were aware of the theory that depression was due to too little serotonin being present at the synapse. This idea came from a variety of studies that included the observation that the drug reserpine caused a depletion of serotonin and could cause depression in some normal individuals. The researchers wanted to find a chemical agent that would block the normal uptake of serotonin at these sites. The approach they developed was to try to synthesise a molecule so similar to serotonin that the neurons would think it actually was, and so it would compete with serotonin for uptake, leaving more serotonin outside the cell. They developed a compound called fluoxetine hydrochloride which was later named Prozac. It took another 13 years of clinical trials on thousands of patients to get the drug approved and licensed.

There has been much discussion about problems associated with antidepressants. The most dramatic are the claims that taking an antidepressant when depressed can precipitate suicide. The evidence suggests that those who are depressed and take an SSRI antidepressant had a twofold increase in the possibility of attempting suicide, but the number of such attempted suicides leading to death was not increased. A possible explanation is that the antidepressant makes the individual agitated and capable of taking the action which, when they were severely depressed, they were just not capable of. Nevertheless it is an aspect that gives antidepressants a very negative image and psychiatrists must be aware of possible suicidal thinking. The other claim is that antidepressants can be addictive and lead to dependence similar to that claimed for Valium. There is just no evidence for such addiction, but giving up antidepressants can present problems as there can be unpleasant withdrawal symptoms. It is usually essential to come off them slowly. There are also many side effects as listed in the table on p. 139.

Fluoxetine is the only antidepressant medication that has Food and Drug Administration (FDA) approval for the treatment of depression in children and adolescents. In June 2003, the FDA issued a statement about possible increased risk of suicidal thinking and suicide attempts in children and adolescents who were treated with Seroxat for major depressive disorder.

The SSRIs act on the serotonin transporter molecules in the neuron cell membrane that move serotonin back into the cell. The various SSRIs differ in their potency to affect uptake but there is no evidence that this has any bearing on their effect on depressed patients. These new drugs are not more potent in terms of their ability to reduce depressive symptoms, but their advantage lies in reduced side effects and so an improved quality of life. Patients find it significantly easier to take drugs like fluoxetine than tricyclics. In several studies the failure to continue taking the drug was less than 10 per cent for fluoxetine and around 30 per cent for the tricyclics.

Depression is often associated with poor and unsatisfactory sexual function. This may reflect both loss of interest in sex itself and pleasure in general. In males impotence is quite common. Both male and female sexual behaviour is affected by serotonin and noradrenaline and antidepressants, particularly the SSRIs, are often associated with sexual dysfunction.

For all these antidepressant drugs there is a problem: why do they take days or weeks to exert their effect on depression when their biological effect occurs within a few hours? The mean time for the onset of action of monoamine oxidase inhibitor is six weeks; for tricyclics and SSRIs three weeks. One would, logically, expect that drugs that inhibit the re-uptake of serotonin would lead to an almost immediate increase in the concentration of serotonin surrounding the neurons. This is not the case. The slow effect of antidepressants that inhibit neurotransmitter uptake makes it clear that their mode of action is not directly related to inhibition of uptake. There is now evidence that a more complex system is involved. A key player is the autoreceptor for serotonin, which can control both the release and synthesis of serotonin and is part of a sensitive feedback system controlling the serotonin neurotransmission (as described in the previous chapter). It could be that in depression the receptors are super-sensitive and that the long-term action of antidepressants is to reduce this sensitivity. A novel approach to improve SSRI antidepressant efficiency is to combine it with a beta-blocker, a family of drugs widely used for anxiety, which binds to the autoreceptor and so blocks its negative feedback function, ensuring that there is no reduction in synthesis of serotonin as its concentration rises. This could result in the SSRIs having a much more rapid effect on depressive symptoms.

While the USA embraced Prozac, Germany has taken a different course and turned to a herbal remedy, St John's Wort (*Hypericum perforatum*), a perennial herb with bright yellow flowers. The flowers are left in vegetable oil for several weeks and the resulting red liquid contains the active ingredient. The herb has been used since Ancient Greek times for a variety of medical complaints and came to be thought of as a way of driving out demons. At the start of the 20th century there are references to it being a remedy for nervous diseases. It is claimed that studies on nearly 2,000 patients have shown it to be a safe and effective treatment for depression. Of particular importance is the fact that there are very few side effects. At present it is only available from shops that deal in herbal medicines.

It is of great significance that an antagonist of substance P (see previous chapter) is a potent antidepressant, for it may not act through serotonin and noradrenaline and so opens up the possibility of the development of a quite new class of antidepressants.

The discovery of the effects of lithium on depression and mania has a curious history. In 1949 an Australian psychiatrist, John Cade, was working on the hypothesis that manic depressives had some chemical in excess while depressed patients lacked something or had too little. This hypothesis was remarkably perceptive given what is now thought about the importance of the concentrations of neurotransmitters. In order to test the possibility that the substance in excess might be secreted in the urine of manic patients he injected some into guinea pigs and found it to be more toxic than urine of normal individuals. In trying to identify the toxic component he used lithium to dissolve one of the compounds being tested and discovered that the lithium reduced the toxic effect. Injecting lithium on its own seemed to tranquillise the guinea pigs. He then tried lithium on manic patients, who responded very positively, but little attention was given to his initial report, and lithium was not used until its action was investigated in what was one of the first proper clinical trials in 1954, when patients were randomly assigned to lithium or a neutral placebo. Its use for manic depression is now almost universal.

TABLE OF ANTIDEPRESSANTS

Drug Name	Trade Name (USA)	Trade Name (UK)
Tricyclics		
Imipramine	Tofranil	Tofranil
Amitripyline	Elavil/Endep	Tryptizol/Lentizol
Nortriptyline	Aventyl	Allegron
Clomipramine	Anafranil	Anafranil
Protriptyline	Vivactil	Concordin
Doxepin	Sinequan	Sinequan
Trimipramine	Surmontil	Surmontil
Dothiepin	Doxepin	Prothiaden
Lofepramine		Gamanil

Side effects include difficulty urinating, constipation, rapid heartbeats, feeling faint when standing, a feeling of drowsiness due to a sedative action, and confusion. There can be effects on heart rhythm. Taking an overdose poses serious risks and can cause death. Anxiety, headache, and tremor have been observed after discontinuing a short course of a tricyclic antidepressant. More severe withdrawal symptoms may occur after taking this class of drug for several years; these may include confusion, nausea, and convulsions.

Drug Name	Trade Name (USA)	Trade Name (UK)
Specific Serotonin Reuptake Inhibitors – SSRIs		
Citalopram		Cipramil
Fluvoxamine	Luvox	Faverin
Fluoxetine	Prozac	Prozac
Paroxetine	Paxil	Seroxat
Sertraline	Zoloft	Lustral
Venlafaxine	Effexor	Efexor

Common adverse effects are nausea, insomnia and general agitation or anxiety; also blocks noradrenaline uptake. There is also evidence of sexual dysfunction. Overdose is, on the whole, not lethal. Withdrawal symptoms can include delirium, nausea, fatigue and dizziness, so the ending of a treatment should take place over a period of several weeks.

Drug Name	Trade Name (USA)	Trade Name (UK)
Monoamine Oxidase Inhibitors – MAOIs		
Phenelzine	Nardil	Nardil
Tranylcypromine	Parnate	Parnate
Moclobemide	Mannerix	

Side effects include lowering of blood pressure leading to dizziness and fainting, headaches and sleep disturbances. Moclobemide, a reversible inhibitor, has the fewest effects and does not require dietary restriction. This class of drugs can however act adversely with particular foods: mature cheese and pickled fish are examples. Abrupt withdrawal can lead to headaches and nightmares.

TABLE OF ANTIDEPRESSANTS (*continued*)

Drug Name	Trade Name (USA)	Trade Name (UK)
Other antidepressants		
Mirtazapine	Remeron	Zispin
Nefazodone	Serzone	Dutonin
Viloxazine		Vivalan
Buproprion	Welbutrin	

Buproprion has the fewest side effects; for the others there are reports of sedation, and in some cases similar effects to the tricyclics.

Drug Name	Trade Name (USA)	Trade Name (UK)
Drugs for manic depression		
Lithium Carbonate	Eskalith/Lithobid	Camcolit/Priadel
Carbamazepine	Tegretol	Tegretol
Sodium valproate	Depakote	Epilim

Lithium has a variety of side effects – a reduction in thyroid function, excessive thirst, a fine tremor of the hands can be observed in a significant number of patients. In early pregnancy lithium can about double the risk of congenital malformations. As some patients develop lithium toxicity in which there is a coarse tremor, nausea, diarrhoea and, in more severe cases, vomiting, its concentration in the blood needs monitoring. Carbamazepine can cause nausea, dizziness, an itchy rash and a lowering of the white blood cell number. With valproate there may be nausea, stomach cramps and, on occasion, liver failure.

Drugs are not the only physical treatments for depression which aim to directly alter brain chemistry. Others include electro-convulsive therapy (ECT), psychosurgery, sleep therapy and light therapy. While the initial response of many people to the first two is one of horror, because they seem so violent and crude, such views should be tempered by the actual evidence.

The history of ECT is not a happy one. It has been marked by intense fear that it can impair mental function, particularly memory, damage bones and the heart and even cause death. The use of electric shocks for alleviating headaches may go back to the first century AD, when an electric eel was placed on the head of a Roman emperor, and Jesuit missionaries reported similar treatments in Ethiopia in the 16th century. In modern times it owes its origin in the 1930s to the idea that convulsion induced by chemicals could be beneficial for some psychiatric patients.

Electrical induction of convulsions was introduced by Italians in Rome in 1938. But its use has until quite recently been fraught with controversy. There are claims, for example, that up to 1977, 384 deaths related to electro-convulsive therapy occurred. In the 1970s in the USA there was even a Network Against Psychiatric Assault which campaigned against the use of ECT. The film *One Flew over the Cuckoo's Nest* encouraged such views since it showed a hapless patient being forcibly treated by an unfeeling and totally unsympathetic staff. It was such pressures that caused the medical profession to lay down guidelines for treatments.

ECT is almost always only given when other treatments have failed and when the patient is very suicidal. It is best given with a brief anaesthesia, a muscle relaxant and oxygen ventilation. Electrodes are placed on either side of the head and current is passed to cause a seizure lasting for about 30 seconds. Such treatments are usually carried out two to three times a week for several weeks. Placing the electrode on one side of the head on the non-dominant hemisphere, as done in Sweden, reduces any effects on memory. The treatment has been described as no worse than a visit to the dentist. There are many reports of memory loss but these are usually only temporary.

Contrary to common sense it is not the applied electricity itself that is the therapeutic agent in ECT but the seizures that it causes. ECT requires the passage of an electric current through the brain sufficient to stimulate the hypothalamus and pituitary and to elicit a grand mal seizure. It is also necessary that the seizures, to be effective, are continued over a month or so. There is no good evidence that ECT exerts its effect in a manner similar to that of drugs which affect neurotransmitters. One view is that ECT exerts its effects via hormones. It could cause the release of hormones from the hypothalamus like ACTH and endorphins and also from the pituitary.

Psychosurgery, not unlike ECT, evokes very frightening images and has had a very bad press. The original operations of prefrontal leucotomy in the 1940s were mainly performed on patients with depression and schizophrenia. While the potential side effects were severe, the surgery at least allowed a significant number of patients to leave hospital. The operation was no longer widely used once drugs for the treatment of schizophrenia became available. In the UK between 1942 and 1954 there were 10,000 such operations and about 50,000 in the USA. The mortality rate was about 3 per cent and 20 per cent were discharged from hospital. About 10 per cent had side effects like epilepsy, and personality changes occurred in 5 per cent of cases. The number of operations now is of the order of hundreds per year world-wide.

It might seem inappropriate to consider brain surgery as a treatment for depression because of the perceived side effects and the distaste felt for operating on so delicate an organ. However, this is to deny the extreme suffering experienced by some depressed patients for whom other treatments have not worked. They might welcome such an attempt to alleviate their condition in much the same way as cancer patients are prepared to take drugs which have serious side effects. For example psychosurgery can be very successful in treating epilepsy. The current techniques make use of lesions induced by radio-frequency in a relatively trauma-free process. The procedure is followed using magnetic resonance imaging and recovery is rapid and full activity can be resumed within 48 hours. Unrelieved depression is still a condition for which psychosurgery ought to be considered.

A new and quite different treatment for depression is deep brain stimulation. Electrodes are inserted in the brain to a specific region and the brain then stimulated, the electrodes being connected to a stimulator attached to the patient. Repeated stimulation of white matter tracts adjacent to the subgenual cingulate gyrus was associated with a striking and sustained remission of depression in four of six patients. Sudden change to a positive mood was already seen when stimulation started in the operating room.

Disturbance in the pattern of sleep is a common symptom of depression. The risk of developing a new depressive episode is much higher with those who have difficulty sleeping. The relationship between sleep disturbance and depression raises several questions: is sleep disturbance a fundamental feature of depression or merely a consequence, and are sleep disorders specific to particular psychiatric conditions or are they similar?

The most common sleep disorders in depressed patients are difficulties in falling asleep, remaining asleep and early morning waking. Another abnormality is in the nature and timing of sleep in which there is Rapid Eye Movement (REM), during which there are often dreams. Normal people have REM sleep several hours after falling asleep and then a period of REM at about hourly intervals. About one half of all depressed patients enter into REM sleep much sooner than normal and they have most of their REM sleep within the first few hours. Antidepressant medications suppress REM sleep and this may play a role in their effect on depression.

A number of sleep manipulations improve the symptoms of depression. These include total and partial sleep deprivation, selective deprivation of REM sleep and changing the pattern of times when the patient is asleep or awake. Total and partial sleep deprivation induces an immediate improvement in some 60 per cent of patients but this disappears when they are allowed to sleep again. Since sleep deprivation in healthy patients is detrimental to many desirable and normal functions such as alertness, fitness and motivation the effect of sleep deprivation in improving the condition of depressed patients is paradoxical. These patients report that they are tired and sleepy but have an increased energy and an improved mood. It is possible that an explanation may be in the circadian rhythm that controls the sleeping/waking cycle and which may be upset in depressed patients. Another possibility is that sleep deprivation might break the psychological deadlock and tension in the patient's brain; it is not clear just what this means, though it sounds nice.

Sleep aside, one should not ignore the periodicity or rhythmicity of certain aspects of depression. It could well be that depression is linked to disorders in the normal circadian rhythms that affect our lives. Reasons to think about such a connection are four features of depression: early morning wakening; variation during the day of the severity of symptoms – typically they are worst in the morning and improve towards evening; the relationship to the seasons as in seasonal affective disorder (SAD); and the long-term cycle of the illness. Many of the hormones and neurotransmitters associated with depression show a circadian rhythm both with respect to release and synthesis. These include serotonin and other neurotransmitter receptors, and the synthesis of serotonin, dopamine, cortisol and related hormones. The peak of cortisol secretion occurs earlier in depressed patients than in normal controls.

Some of the features that distinguish SAD from more classical depression are an increase in sleep and increases in both appetite and weight. There is a craving for carbohydrate-rich foods. And unlike depression, the episodes typically resolve by springtime, though for some individuals they continue into the summer. The condition can also occur in children, who show irritability and can have school problems. Surprisingly, there was little systematic investigation of the condition until the 1980s, though a German psychiatrist had described an engineer whose annual seasonal depression responded well to treatment with a sunlamp. This treatment has been shown to work in a number of clinical trials; however, it has not been easy to control for the placebo effect. Morning seems to be the most effective time for treatment and a 60-minute treatment works well. The eyes seem to mediate the effect of the light, though in one case a blind sufferer responded positively. Unlike drug treatments, there are no reports of side effects on normal subjects exposed to the light regime.

The mechanism by which light treatment works is not known but there are theories which relate it to the suppression of melatonin secretion, affecting serotonin levels and modification of circadian rhythms. Light therapy has also been used to treat patients who have depression not related to seasonal variation. While longer periods of exposure are generally required, a significant number of patients responded positively and so it could be used to help with other treatments. Again, it has the advantage that side effects are minimal.

There is also what is known as summer SAD but it is not nearly as common nor well-researched as its winter companion. Changes in environment like repeated cold showers have not been established as a satisfactory treatment, but patients do respond to antidepressant drugs.

CHAPTER ELEVEN

Psychotherapy

Psychotherapy has its origins in psychoanalytic theory as put forward by Freud. He described the essence of the psychoanalytic method as follows: 'It may be said that the theory of psychoanalysis is an attempt to account for two striking and unexpected facts of observation which emerge whenever an attempt is made to trace the symptoms of a neurotic back to their sources in his past life: the facts of transference and of resistance. Any line of investigation which recognises these two facts and takes them as the starting point of its work may call itself psychoanalysis, though it arrives at results other than my own.'

As many as 200 different types of psychotherapy have been identified which are used in clinical treatment of patients with various psychological disorders. Those for treatments of depression are based on a therapist talking to a patient. These range from short-term psychotherapeutic treatments to much longer treatments like psychoanalysis. Two particular short-term treatments based on theoretical frameworks significantly different from classical psychoanalysis have been used in recent years, interpersonal therapy and cognitive behavioural therapy. Unlike psychoanalysis, cognitive therapy is not devoted to uncovering early unconscious experiences and memories. In practice, however, therapists are often likely to use a combination of treatment strategies.

These new psychotherapeutic treatments of depressed patients are sometimes referred to as supportive psychotherapy, in order to emphasise that they aim to help the patient and make no attempt to remodel their psyche. The therapist aims to gain both information and trust from the patient, and tries to explain the patient's condition as well as trying to find a way of helping the patient get better. In addition, the therapist tries to guide the patient with respect to relationships, work, education and general health. Of particular importance is the necessity to persuade the patient to not make major life changes during their depressed state. The therapist seeks to establish realistic and attainable goals. All this requires considerable skill and approaches must be tailored specially for each patient. Use of medication to help alleviate severe symptoms and make the patient able to respond to psychotherapy is a quite common practice.

Psychoanalytic therapy and some psychodynamic therapies still owe much to Freud's ideas and so, unlike cognitive therapy, do not rest on what can be reasonably regarded as a scientific basis. A central idea is that depression arises from an ambivalent relationship with a lost object and this results in repressed rage which becomes directed against the self. It also results in increased self-criticism and self-destructive impulses. Some psychoanalysts talk of depression being due to a vicious attack from the superego. Psychoanalysts have no hesitation in effectively blaming parents and ignoring any genetic contribution as they believe the cause of depression to be related to an early lack of care, warmth and protection. Depression results from conflicts related to such early deprivations because of repressed fantasies and having excessively high ego ideals. Thus, treatment by psychoanalysis is based on the theory that the psychic processes underlying the illness will continue unless the relevant unconscious forces are brought into consciousness and under the control of the ego.

Transference is claimed to be the unconscious distortion of a relationship to fit expectations of events from the past. Resistance is the attempt to block memories and other aspects of one's inner world from becoming part of one's conscious thoughts. Therapy, for a psychoanalyst, involves undoing resistance and interpreting transference. Self-awareness of one's past is regarded as essential and can be summed up in the maxim that the unexamined life repeats itself. A transference relationship with the therapist can be used to allow detailed examination of hitherto unrecognised patterns of thoughts, feelings and behaviour.

Psychoanalysis, because of its very nature, is not designed to deal directly or rapidly with the symptoms of a severe depression and this is a serious disadvantage. One of the first aims in analysis is to discover what precipitated the depression and this requires the patient to look inward and uncover fantasies, dreams and distortions. However, encouraging the patient to recall distressing past experiences can make the depression worse. Quite often the depressed patient will reinstate with the analyst a relationship they had with an important figure – for example, a parent – from their childhood. The patient will transfer to the analyst values and expectations they have for this figure. A problem can arise if the patient comes to focus their whole existence on the therapist, who thus acquires an unnatural stature. There are accounts, admittedly anecdotal, of depressed patients in analysis who have been advised against medication and other treatments and who have deteriorated over a long period.

While psychoanalysis is traditionally a long-term treatment and tries to restructure the whole of the patient's personality, short-term treatments based on its principles have been devised more recently. These go under the general title of psychodynamic therapies. One of their aims is to restructure the patient's main problems in the context of the transference relationship with the therapist.

Interpersonal therapy is based on the idea that the crucial factor in depression is the social network or interpersonal relationships of the patient. These are in part determined by life events and attachment bonds. For example, Bowlby noted three major common themes in depressed individuals: the experience of never having attained a safe and satisfying relationship with one's parents in spite of numerous attempts; the experience of being repeatedly told that one was unlovable, inadequate or incompetent; or the experience of deprivation following the loss of a parent. These childhood experiences may create irrational belief systems which distort perception and behaviour in adult life. The depressive appears to grow up needing love and approval and believing that their worth is dependent on such approval. The depressive thus strives to get such approval and perverts normal relationships. When an event occurs which strips away such defensive manoeuvres and confirms their belief that they are unloved, then depression results. Even though the evidence for these views is not strong, it is with such issues that many psychotherapies try to deal.

Interpersonal therapy is relatively brief, generally consisting of 15 to 20 sessions. The therapy does not assume that interpersonal problems cause the depression but that depression occurs within an interpersonal context. The aim of the therapy is to understand that context and so to bring about recovery by helping the patient to deal more effectively with relationships. It also helps the patient to understand that the symptoms are part of a well-known illness and that the prognosis is a good one.

Four major areas of interpersonal relationships are commonly associated with major depression – abnormal grief reaction; disputes with other people; inadequate social skills; and difficulties in taking on new roles in relation to life events, such as changing career, graduating from school or university or retiring. Therapy focuses on current, not past, interpersonal relationships. The social context just before and since the onset of depression is all important; however, past episodes of depression and significant relationships have to be taken into account. The patient and the therapist agree, after an initial evaluation, what areas to focus on. While past events and family relationships are assessed, emphasis is on the present, not the past.

Each of the four areas just outlined must be dealt with. For example, in dealing with abnormal reactions to bereavement the approach is to facilitate the mourning process and to allow expression of the feelings, both negative and positive, surrounding the loss. In dealing with interpersonal difficulties, such as those involving a spouse, parent or work colleague, the patient is encouraged to examine distorted perception and faulty communication that make the depression worse. The therapist tries to teach the patient better ways of relating to others, to improve interpersonal skills. For role-transition a strategy is to help the patient to distinguish between real losses and imagined ones and to encourage more realistic goals; new opportunities are emphasised.

A major advance in the treatment of depression comes from the cognitive model of Beck. He maintained that negative thoughts not only characterise depression but are a key factor in maintaining the depressed state. Cognitive therapy aims to alter the patients' thought processes and to teach them new ways of thinking. Beck's theory suggests that people experiencing a depression typically show three related features in their thinking – automatic negative thoughts about the self, the world and the future. These automatic negative thoughts are believed to be sustained by distorted thinking such as selective recall of negative experiences, and the tendency to see things as either wholly bad or good. However, it is still far from clear whether or not negative thoughts are the cause or consequence of depression. In terms of my ideas about malignant sadness I regard them as arising from an abnormal emotional state.

Cognitive behavioural therapists attempt to alter behaviour and cognitive function by altering thinking patterns. They assume that earlier learning experiences are the cause of depression and that the purpose of therapy is to reduce distress and unwanted behaviour by undoing this learning or providing more adaptive responses. It is clear then that the therapy is based on the acceptance of the idea of the pervasive role of unconscious processing in everyday thinking and that this can affect emotions. There are two cognitive systems that are routinely used.

One is automatic and involves processes of which people are not aware and uses implicit memory, and the other is that of which we are conscious, and requires effort and uses explicit (declarative) memories which are accessible. Emotional response may be influenced by both, and both may be influenced by emotion.

It is thus assumed that there is knowledge which is not accessible by conscious processes but which may be retrieved automatically when the appropriate environmental stimulus is present. Therefore, when there is a reminder of an unpleasant event there can be automatic activation of earlier emotions, thoughts, and impulses to behave in a particular way; this could, for example, account for why people sometimes feel inexplicably sad or afraid. Cognitive therapy tries to limit the ease with which these adverse memories are activated. In essence the patient learns new skills to deal with the response and this leads to a change in the patient's beliefs and assumptions about their life.

Cognitive therapy for depression is a brief, structured, directed treatment designed to alter the beliefs and negative thoughts causing the depression. Typically, therapy involves some 20 sessions, or perhaps fewer, over the course of three to four months. A distinctive feature of cognitive therapy is that the therapist regularly asks the patient about any conclusions they have drawn from the discussions, and also any reactions they have had to the therapist. This emphasises the collaborative nature of the therapy. Unlike analysis-based therapies, the therapists do not assume that they have knowledge of the patient's reactions to what occurs in the sessions, and so enquire directly about them rather than offering interpretations.

Another special feature is that sessions usually start by setting out an agenda for the session. This enables discussion to be focused and attention given to the most important issues. Setting an agenda also provides a basis for framing goals and shifting responsibility from the therapist to the patient. In early stages the rationale for cognitive therapy is explained and the patient's own thoughts used to illustrate the method. For example, if the patient felt low and without hope before the session, the therapist would ask what thoughts were associated with these feelings. This can be used to illustrate the relationship between negative thoughts and feelings.

Cognitive therapy also makes use of behavioural techniques, such as positive reinforcement. For example, scheduling of activities particularly helps patients who have become inactive. Such scheduling lays down a detailed plan of action beginning with relatively easy tasks like going for a walk or reading a magazine. One should not underestimate the difficulty some patients might have with such apparently simple tasks. If increased activity then results in an improved mood this is valuable reinforcement for the patient. And even if it does not, this can provide useful feedback for further discussion and planning.

Some patients do not get any pleasure from fulfilling their planned schedule of activities. They may believe that they do not deserve to do things for the sake of their own enjoyment. It is necessary to explain to them that such thinking is self-defeating. In order to help the patient the degree of pleasure and success with each task is monitored by the patient, who rates it on a scale of, say, 0 to 5. Keeping detailed records of this type can prevent the patient denying, in retrospect, any pleasure or success.

A major aspect of cognitive therapy is to teach the patient to monitor and recognise Automatic Negative Thoughts – ANTS is a useful acronym. ANTS are the thoughts going through the patient's mind whenever they are feeling bad. The patient can be asked to monitor, rate and record daily upsetting thoughts and the associated feelings, though some patients find it difficult to distinguish between thoughts and emotions – for example 'I am so upset' is a feeling whereas 'I am disliked at work' is a thought. The therapist then tries to teach the patient techniques for dealing with ANTS. This involves making the patient examine the evidence for the validity of each negative thought. There may indeed be evidence but it is always exaggerated, and predictions from it are unrealistically negative.

Another approach is to ask the patient if there is not another way of looking at their situation. Are there not, perhaps, more plausible interpretations than those claimed by the patient? Explanations can be offered that suggest that change is possible and this can promote both persistence with the therapy and hope. A related technique involves the patient and therapist trying together to find tests for which interpretation of events is the best one. It is important that the patient is encouraged to take action to change upsetting situations. So if a patient says 'no one cares for me' it may well be true that the patient has in fact been rejected. Focus must then be on training the patient in social skills and changing the perception and expectation that the situation is hopeless and cannot get better.

Dealing with negative thoughts is at the core of cognitive therapy. A number of techniques are used that include direct questions, the use of mental imagery, role playing, keeping a diary and using moments of strong emotion to gain access to the negative thoughts. Some examples of these techniques follow and are taken from Blackburn and Davidson's book *Cognitive Therapy for Depression and Anxiety*.

Underlying many of the ANTS are a set of beliefs and it is one of the therapist's tasks to uncover these beliefs. One way is to keep asking 'What would be so upsetting to you about that?' until a possible basic belief has been revealed. In order to help the patient come to terms with their negative thoughts the therapist can ask direct questions such as 'What was going through your mind then?' Some patients have difficulty answering such questions and use is then made of 'guided discovery' or inductive questioning, which tries to get the patient to recreate what was being thought.

PATIENT: I felt terribly upset yesterday when I came back from work and I do not even know why.

THERAPIST: What was going through your mind at the time? (*direct question*)

P: I don't know. Nothing in particular. I just felt this black cloud come over me.

T: Was this before you got home or after you got home?

P: I think I was beginning to be upset before I left work. But it just got worse and worse.

T: Had something happened at work?

P: Nothing much that I can remember. Nothing out of the ordinary, anyway.
T: What do you mean by 'nothing out of the ordinary'?
P: I was just doing my usual work; there were no classes to teach and I was marking some papers.
T: Is this when you started feeling bad?
P: Yes, that's right.
T: Did anybody come into your office?
P: No, nobody came.
T: Was there any interruption, like a telephone call?
P: No, nothing at all.
T: Did it bother you that nobody called or telephoned?
P: No, I was relieved not to be interrupted for once and to be able to get on with the work.
T: While you were marking the papers, were you able to concentrate all the time?
P: No, you know what it's like. It's the same when I watch television. My mind is not really on it.
T: Did you have thoughts or images going through your mind?
P: Yes, I suppose so. I was beginning to think about going home.
T: So here you were – sitting at your desk and it's getting near the time to go home. Is this what went through your mind? Did you have an image of your home? (*Therapist tries to create a concrete image.*)
P: Hmm. Hmm.
T: Can you tell me what the image was?
P: Yes . . . I was thinking about the house being cold and no one being there – sitting by myself and forcing myself to eat some supper and the telephone not ringing. (*Patient starts crying.*)
T: OK – you created a very sad image in your mind and this made you feel depressed. You seem to have been painting a very black picture which made you feel low. Now, you could have had a very different image of going home. Let's try. Let us say you had a picture in your mind of going home, putting on the fire, making yourself a nice supper whilst listening to the radio and sitting down for the evening in front of a warm fire, watching a good film on TV. You phone a friend and have an interesting chat. How do you think this picture would have made you feel?
P: I guess I would not have felt so low.
T: Yes, it is sometimes difficult to trace the pictures that go through our mind which colour our mood. Here, I think you have succeeded in tracing what started your black mood. It seems to me that what you did was jump to conclusions about how your evening would turn out and you managed to persuade yourself that the image was a reality. Do you see that?
P: Yes, maybe I did get carried away there.

T: Now let us see what is so depressing for you about going home to an empty house.

Another technique tries to make use of mental imagery:

T: Try to recreate exactly what happened on Wednesday night. You were at home with your husband. It was after supper and you had put the children to bed. You were in the living room, watching television and your husband was reading the newspaper. Close your eyes and try to imagine the situation in as many details as possible, the sitting room where you were sitting, the time of day, and so on.

T: (*two minutes later*) OK. Have you got the picture in your mind?

P: Yes, I see it now. I had gone to sit close to John on the sofa and he was reading the paper. After a while, he got up, turned the volume of the television down a bit and sat somewhere else quite far from me.

T: Did you attach some meaning to that?

P: Yes, I thought 'He can't stand being near me. He finds me boring. He does not love me any more'.

T: Well done. We've now got hold of the hot cognition, as it's sometimes called. Do you now see why you felt so hopeless and desperate? Your husband behaved in a certain way and you put quite drastic interpretations on his behaviour. Somehow, what he did had something to do with you. You personalized his behaviour and somehow it all reflected badly on you. Let's look at these thoughts again, to see how realistic they are.

It is important to help the patient examine the evidence for their negative thoughts:

T: So, you feel anxious every morning because you think you won't be able to cope, that you will collapse and have to be taken home.

P: Yes, every morning it's the same. The moment I wake up, I feel a knot in my stomach. I have to force myself to get to school.

T: What goes through your mind at these times?

P: Oh . . . 30 unruly 7-year-olds, calls from an irate headmistress, the lessons not properly prepared, not interesting enough . . .

T: OK, you imagine or even predict that a whole lot of disastrous events are going to take place and that you won't be able to cope?

P: That's it. I really can't stand it any more.

T: Has any of these events taken place in reality recently?

P: Oh yes. Ever since I was moved to a younger age group a year ago.

T: Have you ever been unable to cope and had to be sent home?

P: No, but it nearly happened a few times.

T: It nearly did, but it's important to note that it did not. So, what is it that you cannot stand, the images which come to your mind in the morning or the reality of the classroom?

P: Well, strangely, I don't get so anxious in the classroom. The images are worse than the reality.

T: Exactly. You say to yourself 'I won't be able to cope, I will collapse'. But, if I understand correctly, difficult situations have occurred and you have not been sent home. So your predictions are constantly invalidated, but you don't believe the evidence for some reason. Is that right?

P: That's right. You would not believe that I trained as a scientist would you?

T: Well, that's what this therapy is about, to try and help you be a scientist in your own life.

The therapist can suggest alternative interpretations:

T: So, what's happening is that you are facing a number of new situations at the moment and these make you anxious. Is that right?

P: Everything keeps changing and I don't like that.

T: OK, you don't like new situations, but you have coped with them without major disasters over the last year?

P: Yes.

T: Are situations less novel now than they were 9 months ago?

P: Yes, and it is getting easier. These 7-year-olds are not so bad most of the time.

T: Good. So, if you said this to yourself in the mornings instead, would it be easier? Perhaps something like: 'This is still a relatively new situation for me, but I'm coping with it. The disasters that I fear have not occurred and are less likely to happen as I get more familiar with the new class and the new headmistress. I'm doing not too badly really'.

In spite of the different approaches, therapists respond to the style of their patient. If, for example, they see that the patient is over-involved in personal relationships then they tend to use the techniques of interpersonal therapy, whereas if personal relations are not that significant they resort more to the techniques of cognitive therapy. A common factor in the success of therapy is thought to be due to what is called the therapeutic alliance. This involves the patient's belief that the therapy will help and the therapist helping the patient to learn something, to gain insight. Therapists thus intervene because they think it will be helpful and they adjust their ideas from moment to moment in response to the patient's behaviour.

One view of how psychotherapy works is by a process of what is known as assimilation. In this process the patient moves from being unaware of unwanted thoughts to experiencing them as painful, then less disturbing and gradually merely puzzling. Finally they are understood and mastered. Perhaps psychotherapy works to break the positive feedback loop of sadness driving, and being driven by, negative thoughts by dampening down those thoughts.

But how well does psychotherapy compare with drug-based treatments?

CHAPTER TWELVE

What Works?

My waking up one morning when I was in the early stages of my depression with the overwhelming desire to commit suicide resulted in frantic calls to doctors and I was admitted to the psychiatric ward of the local hospital. For reasons I have never fully understood being in hospital seemed essential for my recovery; William Styron too, has written about the value to him of being in hospital. Perhaps we needed the security that it provided. Initially I was not allowed out of the ward on my own but since I was a voluntary patient I persuaded them that if I really wanted to kill myself I would, and so they allowed me the freedom to come and go during the day.

I had stopped taking the tricyclic drugs that had me even more anxious and was now put onto another antidepressant, one of Prozac's first cousins, Seroxat, an SSRI drug. When I asked my psychiatrist why she did not put me on Prozac she explained that in her experience her patients tolerated Seroxat better. She was also extremely reassuring, telling me again and again that depression is self-limiting and that I would recover. I did not believe a single word. It was inconceivable to me that I should ever recover. The idea that I might be well enough to work again was unimaginable and I cancelled commitments months ahead. I felt particularly guilty towards my wife and my children who had to look after me each day when I left the ward. I could not be left on my own for even a few hours as I was too anxious. I would look out of the hospital window and see the courts where

I had regularly played tennis and was filled with despair fed by the conviction that I would never play again. I thought I would end up a bedridden zombie.

I was taken out every morning by one of my family and returned around five each evening. I had been told how to do relaxation exercises which helped a bit. I was to lie on my bed with my eyes closed and relax each muscle in turn starting with my back until even my face muscles were not tense. When I was conscious only of my own breathing I was to imagine myself in a beautiful place. One of the psychiatrists tried to give me cognitive therapy but the technique at that stage was out of my grasp. I could just recognise how negative I was but it seemed perfectly rational in the condition I was in. I kept asking for ECT as I was desperate and did not believe the drugs I was being given would work. I was completely negative and self-involved. I had no emotions, no feelings and was unable to cry, but curiously retained a macabre sense of humour. Nothing gave me pleasure and every decision, no matter how small, increased my anxiety. The days were interminable and I wanted to remain curled up in bed all day, though by evening I began to feel better and could watch TV and read. But the nights were terrible as I could not sleep properly. The staff were extremely supportive and helped me get through the night with doses of sleeping pills that left me dopey the next day. I would wake in the early morning feeling that my skin was on fire and desperate to go back to sleep, but sleeping pills were not allowed after 3 a.m. Each morning I was back in the black pit. 'Classical,' I was told, which was no consolation.

I was, I believed, not so much depressed as excessively anxious – they unpersuasively claimed that depression and anxiety were really very similar – and had repeated panic attacks. I convinced myself with an impeccable private logic that my arrhythmia was uncontrollable and also that I had Parkinson's disease since I developed a noticeable tremor, the tea cup clattering in its saucer as I carried it. I also heard my foot flapping as I walked and the hospital actually diagnosed a muscle weakness, possibly as a result of a bicycle accident in my past. It was essential to me that

I convince everyone of the hopelessness of my case, and I was acutely irritated that they would not agree with me. My thought processes were often confused and I could think of nothing but my own terrible condition. My memory seemed to be failing badly. On occasion I felt that time had stopped. I was very frightened that I was going insane. Being in hospital was both shocking and comforting. Sometimes I wondered what I was doing living with these old sick people; yet I also felt I was probably the sickest of them all. Thoughts of suicide rarely left me.

However, after several tortuous weeks I began slowly to improve. I could leave the hospital on my own and go for walks. There were bad days and better days and – importantly – I could tell the difference. I began to think about leaving hospital, yet was anxious at the prospect of having to cope at home. I spent hours imagining the impossibility of my heart's potential arrhythmia coming under control and was terrified at the prospect of taking a new drug for it. My hand was always on my wrist, taking my pulse. Events forced me out. By a remarkable coincidence an old friend was admitted to the ward and placed in my room. He snored like a chainsaw – I left hospital sooner than I should have.

Being at home put a great strain on my wife for I found it very difficult being left on my own in the flat for even quite short periods, even a few hours. I started cognitive therapy as an outpatient and this was tremendously helpful. At our first session I was quite hysterical about taking a new anti-arrhythmic drug which carried a warning that if one was exposed to direct sunlight one might turn blue irreversibly. My therapist phoned my cardiologist who allowed me to delay taking it. But there was a little courage left and the next day I did begin taking the new drug, though I went outside covered with sun-cream and in long sleeves even though I was in London in late April.

My therapist introduced me to a new set of relaxation techniques which she suggested I do each time anxiety or panic attacks started. These involved tightening and the relaxing each muscle in turn and I was, if necessary, to spend the whole day doing them. She reassured me that I would not go mad. Cognitive therapy is for me, thankfully quite unlike psychoanalysis whose practitioners are in my experience frighteningly overconfident, almost omnipotent in their convictions. Instead of the therapist probing the inner working of my psyche in order to discover three more people – the ego, id and superego, who cohabit unhappily – cognitive therapy is pragmatic and aims to alter thought processes and to deal with one's persistent negative thoughts.

My therapist also gave me invaluable advice on how to get to sleep – change *black* into *white* one letter at a time. A little bit like counting sheep, the technique is to fill the mind with irrelevant problems. It worked. I also found searching for prime numbers helpful. Best of all was recalling the register, alphabetically, of all those in my class at school – I seldom reached my own name. All this allowed me to start cutting down on sleeping pills by carefully shaving them smaller each night over a period of weeks.

Cognitive therapy may sometimes sound like little more than common sense, but I really needed it and it helped. As I slowly improved, I contemplated going to my first committee meeting since falling ill. What my therapist helped me envisage were the realities of the likely negative possibilities; how bad would it be, for example, if I went and then had to leave? Would my colleagues be very critical? I decided to go. Moreover, that afternoon I cycled to work for the first time since becoming ill. Although I contributed little, the committee meeting went well and marked a major step in my recovery.

I may also have been assisted in the early stages by what my wife Jill called my media cure. The *Today* radio programme phoned me one evening and asked me if I would comment on the first working-out of the complete DNA sequence of a cell. I agreed and their radio car came round the next morning at about 7.30 a.m. – the first time I had been functional that early for months. The interview went well.

With a few setbacks I slowly returned to work and over a

period of about a month gradually came off both antidepressants and sleeping pills. It was miraculous to me that I returned to normal: at the time I felt like Lazarus risen from the dead and given a second chance. Or rather, not quite normal; for I was mildly more manic for a while, a little more reckless and, unsurprisingly, very interested in every aspect of depression.

No doubt both the antidepressants and the cognitive therapy were essential. But this is no more than one story of a recovery from a severe depression. There are many thousands of others. I believe that I was helped by the medication and the therapy, but is it true? The only way to find out what works for depression is by evidence-based medicine.

Doctors are under great pressure and have enormous responsibilities. Mental as well as other illnesses can be complicated and difficult to diagnose and treat. How are doctors to be kept at the cutting edge of new advances? The most reliable way to assess any medical treatment is by random clinical trials, preferably double blind. In such trials treatment is given to some patients and not to others; the latter act as the control group. The patients are assigned to one or other treatment randomly, so neither the patients nor the doctors should know who has been given which treatment. This anonymity is essential for if either group knows what is going on, this can significantly affect the outcome in all sorts of subtle ways.

Such trials are a relatively recent procedure. Trials of any sort to find out if a medical treatment works were virtually unknown until the pioneering work of Pierre Louis in Paris in the 1830s; he showed, at last, that bloodletting on the basis of the ancient belief that the patients had too much of a particular humour did them harm, not good. The medical profession at the time responded with hostility on the grounds that such trials ignored the individuality of each patient, which they alone could take into account when prescribing treatment. Such a misguided holistic stance remains typical of alternative medicine today. An honourable exception to the medical profession's failure to carry out trials is that of James Lind in 1747, who tested the standard treatment for scurvy, sulphuric acid, and found it to be useless. But there was no further reference to this trial for at least 150 years.

Now the situation is very different, for roughly one million randomised clinical trials in all areas of medicine have been conducted over the last 50 years. Clinicians need new and important information far more often than they realise; studies show that in a typical day as many as eight decisions would have been different if the necessary information had been available. Their problem is that they just do not have the time to go searching and the textbooks are too often out of date and the journals too user-unfriendly. One might have thought that refresher courses and their like could provide the solution, but a randomised controlled trial of refresher courses showed that this method just does not work.

The aims of treatments for depression are not only to improve mood and general condition, but to improve social functioning and prevent recurrence. So what works and for whom? A central question is the relative effectiveness of different treatments. Are some drugs more effective than others? What are the side effects? How does one select a drug for a particular patient? Are drugs more effective than psychotherapy? The answers lie in clinical trials but these are expensive and have to be very carefully designed. For example, it is necessary to ensure that those on placebo or medication do not know which treatment they have been given by, for example, experiencing side effects. Also, how is success of the treatment for depression to be judged? Success of the treatment can be based on the absence of certain symptoms or the use of the Beck or Hamilton measures of depression, but what reduction in score is to be taken as a significant improvement or remission?

The terminology used to describe the course of depression makes use of a generally accepted set of terms. One speaks of an individual 'responding' to treatment and thus entering a state of 'remission'. If this lasts a sufficient time – not specified, but a year is not an unreasonable estimate – the patient is said to have 'recovered'. However, if in the state of remission a depression occurs then this is termed a 'relapse' as distinct from a 'recurrence', which refers to a depressive episode following a period of recovery. A 'chronic' depression is one that persists for at least two years and this occurs in around 15 per cent of all patients.

Practising evidence-based medicine for the treatment of depression requires doctors to adopt an evidence-oriented attitude and to become more determined to seek the relevant information. Too often doctors or therapists treating depression argue for the effectiveness of a treatment that they have been using on the basis of the small group of patients that they have been looking after; this is totally unacceptable as the only respectable method is trials with controls. Just consider a therapy for colds based on, say, singing the national anthem before breakfast: the evidence that the patients eventually recovered would be overwhelming, but it would have had nothing to do with the singing. Controls who did not sing would be essential, and in this case would recover just as well.

It is essential to recognise the extent to which patients can recover on their own and it is claimed that 80 per cent of all depressions will eventually undergo remission without treatment, though this may take a year or longer. The placebo effect can also be surprisingly powerful; the patient's belief that the treatment can work even though the 'dummy' medication they are taking contains no active ingredient, or the very act of talking to a psychiatrist, can, in trial after trial, give a positive response of around 30 to 40 per cent. The placebo effect as measured in trials is made up of several elements which are very difficult to disentangle: natural remission, spontaneous fluctuations and the true placebo effect. A further complicating factor in these trials is that as many as one third of those taking part may drop out of the trial, and yet another problem is patient compliance; they often do not take their medication as directed. One more factor to be taken into account is the interpretation of the results of the trial, including an assessment of how well it has been designed; there is a tendency for those with particular views, say in favour of medication or psychotherapy, to bias their interpretation in their preferred direction.

In a review of over 70 studies in 1993 it was found that two thirds of patients given antidepressant drugs responded, in comparison to one third who responded to a placebo. That the placebo treatment had so large an effect is in itself striking, as is the fact that about one third of those in the trial did not respond at all. Also the criteria for a positive response affect the success of treatments; if one uses the criterion that there are no depressive symptoms for two months then the response rate falls to just more than a third.

Results show that patients receiving psychotherapy do about as well as those treated with drugs, though patients with a severe depression initially do better when treated with drugs. A major research programme has been carried out by the National Institute of Mental Health in the USA to compare the efficacy of cognitive therapy and interpersonal therapy and medication for treating depression. This study is regarded as the best of its kind. Research was carried out at three research sites and 239 patients who were moderately to severely depressed entered the trial. Of these 60 per cent had been depressed for more than 6 months and for the others this was the first episode.

Patients in the trial were randomly assigned to four different treatments: cognitive therapy; interpersonal therapy; tricyclic (imipramine) medication; or placebo. The two groups on the medication and placebo had 20-minute meetings once a week with a physician to discuss their medication and general condition and even receive advice. Thus, these groups had some minimal psychotherapy. The therapeutic sessions were taped and every effort was made to ensure that the therapy was according to the appropriate psychotherapeutic procedure. Patients were assessed at monthly intervals for 4 months and then followed up at 6, 12 and 18 months. The assessment used standardised measures such as the Beck Depression Inventory. All treatments resulted in a significant improvement including the placebo group with weekly consultations. The drop-out rate for all treatments was around one third. There was a high rate of relapse after 18 months. Another study compared cognitive therapy with interpersonal therapy and again no significant differences could be detected though both were effective. There was even some evidence that as few as eight therapeutic sessions could be effective.

While different trials sometimes show that drugs are slightly better than psychotherapy, or the reverse, and that some treatments seem better for mild depression while others seem better for severe depression, the overwhelming impression is that the results are best summed up by what one psychiatrist, quoting from *Alice in Wonderland*, has called 'the Dodo hypothesis'. The Dodo after watching a race proclaimed: 'Everyone has won, and all must have prizes.' Taking all the results together it is hard to avoid the conclusion that there is little significant difference in efficacy between any of the current treatments, either medication or psychotherapy, although the side effects can vary considerably. About two thirds of patients will improve and one third will not. But more than one third would improve with a placebo or no treatment at all. This means that in order to achieve one additional 'cure' about three patients will have to be treated. The pressing and serious problem is that there is, at present, no reliable way of knowing which treatment to give to a particular patient. In more general terms one can think of both antidepressants and psychotherapy as working by breaking the loop in which sadness and negative thinking reinforce each other.

There is no question that antidepressant drugs can reduce depressive disorders of all levels of severity in the short term and reduce recurrence in the long term. One patient commented that 'antidepressants have made me feel less depressed when I have carried on taking them. The only problem is that they take a few weeks to start working. They also make you feel as though you're taking steps to make yourself well.' But in spite of the variety of drugs available, numerous studies have not shown them to have any significant differences in their effectiveness but they do have different adverse effects; as another patient commented, 'I find the side effects to be damaging. Felt feelings of fatigue and drowsiness which made me zombie-like.' The SSRIs seem to be more acceptable to patients than the tricyclics, judging by the drop-out rate in trials, and while they are more expensive the lower drop-out rate makes the cost per patient helped similar to that for patients on tricyclics. Monoamine oxidase inhibitors are less effective with severe depressions but may help in special cases. St John's Wort is effective for mild to moderate depression and there are fewer than half the number of side effects. All drug treatments may be enhanced by intensive patient education and the general practitioner working closely with the psychiatrist. Indeed a combination of antidepressant medication and psychotherapy can be very helpful.

An important practical issue is the level of the antidepressant drug in the patient's circulation as there is wide variation in the extent to which individual patient's cells break down the drug. If the concentration is too low then a poor response is to be expected, but if too high there may be adverse side effects. Nevertheless, this area is one which is also poorly understood and there is no evidence that the concentration in the blood of SSRI antidepressants is related to response. The delay in response to antidepressants is typically three weeks for tricyclics and SSRIs and six weeks for monoamine oxidase inhibitors but even these figures are disputed and a positive effect may occur sooner than generally thought. The average time taken to become symptom-free is about 13 weeks.

Despite their success for adults, there is little evidence that antidepressant drugs, particularly tricyclics, work on children. However, there is some indication that SSRIs do lead to an improved response compared to placebo. Cognitive therapy does yield positive results with children, as does interpersonal therapy.

Psychotherapy too has been very beneficial and many patients like it; one patient said, 'It taught me coping strategies and looking at life's situations in a new and more constructive way. Also taught me not to keep pushing myself all the time.' Compared to medication side effects of psychotherapy are minimal but there are cases where the patient does not like the treatment. Two patients' comments: 'I eventually learnt that individual psychotherapy was totally unsuitable for my condition. The therapist left me with problems I didn't think I had before – a self-fulfilling prophecy!' and 'although the therapy was not very damaging, at the time I found it very difficult to handle. Too many issues from my abusive past came tumbling out all at once, and I found myself at the end of each one-hour session with a sense of serious disintegration.' Patients who do not respond to psychotherapy in six to eight weeks should probably be put on medication, and patients with very severe depression should not be treated by psychotherapy alone. Patient preference can also play an important role.

In assessing quite different psychotherapeutic techniques, not just cognitive and interpersonal therapies, what is known as the 'equivalence paradox' has emerged, and this supports the Dodo hypothesis. Treatments in which the transactions between patient and therapist are supposed to be dissimilar are apparently equally effective. The resolution of this paradox is far from clear. One of the problems with research in this area is that psychological therapy requires active participation by the patient. The skill of a good therapist may lie in the ability to detect how best to deal with a particular patient and so establish what is called the therapeutic alliance, which promotes a successful outcome. This ability can help to confound differences between different therapies. There is even evidence that some patients benefit from studying one of the many self-help books on the subject. However, some psychotherapeutic therapeutic approaches, particularly those based on psychoanalytic principles, have not yet been shown to be effective; a counter-argument is that just because the research has not been done, this is not to be taken as evidence for ineffectiveness – not a very persuasive argument.

Patients with manic depression are helped by lithium and, like the other drugs, are some 20 to 30 per cent better than with a placebo. One patient commented, 'Since stabilising on lithium I have not been dotty. "Helpful" therefore is an understatement, for my use of lithium gave me sanity.' But those taking lithium are not always satisfied: 'I feel trapped by having said it was helpful. It is a drug and I hate it. It has not really helped me but right now I need to take it as there is no other alternative.' Manic-depressive patients quite often miss their highs and so stop taking their medication. Depressive episodes of manic-depressive patients respond to antidepressants to the same extent as those with a major depressive disorder. There is also some evidence that cognitive therapy can help.

Anyone who has been depressed and then recovered will always remain aware of the possibility of the depression coming back. It is as if there is a dark cloud somewhere, waiting, ready to descend. I am always conscious of signs of the cloud returning and try to do something to prevent it. The fear of a depression returning can be very frightening. When my wife was dying of cancer within two years of my depression, neither of us was actually depressed. It was a shared, conscious decision to try to be positive, but I believe that I was helped by my regular exercise, particularly jogging. It was when she died that I had feelings that I was once again entering into a depression. The slightest difficulties, like toothache, set me along a pathway of severe anxiety. I partly dealt with the danger by being excessively busy, even manic. I was particularly helped by my daughter pointing out that I was confusing the feelings of normal mourning with those of depression.

Prevention of relapse or recurrence is in fact a problem of major concern. About one quarter of patients relapse within a year, and three quarters will have a recurrence within ten years. Only about one in ten can expect to remain well without a further depressive episode. It is such relapse or recurrence that suggests that a single depressive episode is more serious in the long term than generally recognised, as it may indicate a lifelong problem. In a comprehensive review of many trials it was concluded that brief periods of cognitive and interpersonal therapy are very effective in the short term but their long-term ability to prevent relapse is far from clear. Long-term medication – that is for about a year – may be the best way at present of preventing relapse and it helps to combine it with psychotherapy.

The more severe the initial depression and the greater the number of depressive episodes, the greater the probability of relapse. It may be sensible to continue treatment until all depressive symptoms, or at least most of them, have disappeared. The relapse rate was twice as high for those who discontinued medication or psychotherapy. These results emphasise that it may be wise, even necessary, to continue medication for at least 6 months after the initial positive response to treatment of a severe depression, and this is also true for psychotherapy. There is in addition, always the risk of suicide, and around 10 to 15 per cent of depressed patients will eventually die by suicide, while probably around twice that number will make unsuccessful attempts.

There are a number of patients, around 20 per cent, who do not respond to what is considered to be adequate treatment. While what is really adequate is debatable, such patients are classified as having resistant depression, particularly if they have failed to respond to two successive courses of antidepressant medication using different drugs. This could reflect an inadequate dose or an inadequate length of treatment. There is also the difficulty of the criteria used for deciding whether or not the treatment has worked. One measure is that a fall in the Hamilton Depression Rating Scale by one half is a positive response. But for patients who start with very severe depression such an improvement can still leave them with serious symptoms. Switching drugs can sometimes be helpful and if neither tricyclics nor SSRIs work, the monoamine oxidase inhibitors might. Combining two drugs at the same time runs the risk of their interacting adversely. Lithium too can help and is a major treatment of cases of resistant depression. Addition of lithium to an antidepressant is beneficial in about half of the cases, but it takes up to six weeks for the effect to show itself. In some cases steroids can help.

In very severe cases of depression, ECT or psychosurgery may be necessary. The stigma associated with electro-convulsive therapy has little justification. Its value in relieving both depressive mood disorders and mania is well documented and confirmed by double blind clinical trials. In 1995 the number of patients treated in the USA was around 40,000, which is similar to the number receiving coronary bypass surgery. In Sweden it is regarded as the treatment for depression with the fewest side effects and is routinely used in outpatient clinics. Its disadvantage is that the relapse rate is high though the initial response is very good. Who then should receive this treatment? Clearly those who are suicidal or with very severe depression should be considered, as one virtue of ECT is its rapid response. The best predictor of a good response with ECT is the number of severe depressive symptoms such as hallucinations and inability to move properly.

A major difficulty in evaluating psychosurgery is the absence of proper controlled trials; these would, ethically, be difficult to carry out, for the control would require withholding treatment from severely ill patients. Moreover the numbers are very small. Depression is still a condition for which psychosurgery ought to be considered in extreme circumstances.

Elizabeth Wurtzel was greatly helped by Prozac yet she poignantly puts the role of antidepressant drugs into a more personal perspective. 'The secret I sometimes think that only I know is that Prozac really isn't that great. Of course I can say this and still believe that Prozac was the miracle that saved my life and jump-started me out of a full-time state of depression – which would probably seem to most people reason enough to think of the drug as manna from heaven. But after six years on Prozac, I know that it is not the end but the beginning. Mental health is so much more complicated than any pill that any mortal could invent.'

CHAPTER THIRTEEN

An Excursion to the East

There has also been a strong movement to impose Western criteria and diagnosis of mental illness on diverse cultures, cultures whose traditions and modes of thought are very different from those in the West (as discussed in Chapter 4). One view widely held by medical anthropologists is that in other cultures depression is not experienced as sadness, hopelessness or guilt, but is somatised as bodily symptoms like headaches and stomach pains. It seemed hard for me to believe that depression did not have a common biological basis so I took the opportunity to try to find out for myself, and set off to talk to psychiatrists in cultures very different from our own. I wanted to know if they thought that depression is a universal experience and whether there were other ways of treating it.

Professor Saxena is a psychiatrist at a large general hospital, research institute and medical school in Delhi. There is a high service load, the focus being on outpatient treatment, with 6,000 patients being seen each day. Only 10 rupees (15p) is charged for the first visit, and the poor are treated for free. Saxena claimed that 40 per cent of those attending as outpatients are depressed in one form or another. The criteria used for making the diagnosis are the same as those in the West, though there is less guilt and more somatisation: a burning skin, or a feeling of a fire inside, particularly the stomach. Treatment is by drugs, particularly tricyclics like imipramine, and this seems to work well. Cognitive therapy, or any other psychotherapy, is hardly used.

The orientation is more on the family and their role in providing social support. Indeed, when it is necessary for a patient to be admitted to hospital, a member of the family always accompanies them and a family member remains with them for the several weeks usually required for their recovery. This not only has the advantage that the patient has someone to look after them but also that the family learns to understand the patient and the illness. Therapy thus deals with the immediate problems, particularly relationships within the family, and so the family is often both the cause and the cure: there is a stigma attached to mental illness in India and the family of someone who is depressed experience both shame and guilt. Chronic life events like family problems are more important in causing depression than acute life events. Death is more accepted than in the West and bereavement is not a major problem. ECT is regarded as safe and used with patients who do not respond to drugs or are suicidal. Suicide is quite common among untreated patients.

Many patients also make use of traditional Indian medical systems like the Ayurvedic, which is based on herbal drugs, as well as more spirit-based practitioners like faith healers. The psychiatrists respect their patients' use of other systems of treatment. It is felt that by being too scientific one might be ignoring important needs of the patients. Religion, too, plays an important role in Indian life and religious beliefs influence the presentation of symptoms. Some patients feel they are ordained to suffer because of God's will. But this can mean that they are not responsible for their condition and so feel less guilt and shame. Thus one can try to appease the 'Gods' and the outlook is optimistic. There is no Hindi word for depression, but of course there are words for feeling sad, or having low energy.

At the National Institute for Mental Health in Bangalore, Professor Raguram was sure that depression was a universal human illness and that while there was much emphasis in Western countries on somatisation in countries like India, this was to ignore just how much somatisation there was in the West itself. He thought of depression in terms of it being severe sadness and pointed out that there are good descriptions of such severe sadness in classical Indian literature. But severe sadness is not generally regarded as a medical condition and is often treated by folk remedies. Fate is always the first explanation that is offered. Only when it becomes disabling might the individual end up consulting a psychiatrist. One problem is that an episode of depression always leaves an anxious residue that makes the patients watchful for its recurrence. This constant self-monitoring can lead to a self-fulfilling result, since the family, too, are also watching, perhaps too attentively, for signs of a relapse.

I was struck by the many similarities between the somatisation of depression as experienced in India and my experiences in Peru and Turkey. For example, a psychiatrist in Lima told me that patients typically had pains in their stomach and they went from doctor to doctor, both conventional and folk healers, and only rarely were referred to a psychiatrist. However, when they were referred to him, he often could help them with antidepressants and the symptoms would disappear. As in India, patients in Turkey did not like being told they were depressed.

Professor Kapur at the National Institute of Advanced Studies in Bangalore is convinced that India's rich culture has more to offer in its treatment of depression than just aping Western treatments. I visited him at his small but very comfortable home near the hospital and was very surprised to find that he used yoga in treating his patients.

His experience of working in rural areas where the society was changing from matrilinear to patrilinear has convinced him of the importance of stability for mental health. Also, though illiterate, the villagers can express their problems extremely well. They usually shared their problems with the local healer who could deal with a wide range of problems that might include why the cows were not giving milk. Healers in India probably deal with at least half of the cases of depression. While somatic symptoms are common, Kapur thinks the underlying feelings of those who are depressed are the same everywhere. While accepting that there may be an underlying biological cause predisposing an individual to depression, he finds this theory unhelpful when treating his patients.

Kapur decided that he wanted to find out, as a clinician, what yoga means to those who practise it. He then obtained a grant from the government's Indian Medical Research council to allow him to study yoga for a year, a possibility unthinkable in almost any Western society. He learned the meaning of the mind becoming quiet and so enabling other things to happen. He found through yoga that he was less driven, slept better and ate less. He now uses yoga in his treatment of patients and finds that breathing exercises are helpful in treating depression – it seems to enable the cloud to lift. For anxiety, meditation is his preferred treatment.

Dr Yutaka Ono in Tokyo thinks of loss of interpersonal relationships as a key feature in causing depression. In about half of his patients the symptoms are physical – headaches and abdominal pain being common. He has no hesitation in using antidepressive drugs like the tricyclics. Prozac-related drugs are in his opinion less successful and are little used in Japan, partly as a result of the difficulties in getting new drugs accepted in Japan. Being diagnosed as having a depressive illness is regarded by the patients' families as shameful, though articles on depression have begun to appear in the newspapers.

In his practice Ono uses a type of psychotherapy based on Buddhism, which is not a mainstream practice in Japan. A special feature of his treatment is to combine psychotherapy with Zen Buddhism, though he does not tell his patients he is using a Zen approach. He is acutely conscious of the short amount of time psychiatrists have to spend with their patients – 15 minutes is typical because of the heavy demand on their services – and it is for this reason Zen has been introduced. The aim is to create positive expectations together with therapy as a corrective emotional experience. A key concept is *ichgoichie*, from the Japanese tea ceremony, in which the host is supposed to think that this will be his only opportunity to entertain his guest, and the psychiatrist must similarly try hard to promote a meaningful psychological interaction with the patient. Again, similarly to the tea ceremony, the doctor should, before seeing the next patient, reflect silently on the patient who has just left.

Silence, a key concept in Zen Buddhism, is used by the doctor to connect with the patient. It is not an easy concept. 'Silence continuously exists from the past through to the future and it creates the communication medium in which we exist', and includes nonverbal interchange. The idea of helping oneself overcome helplessness through self-empowerment has a long history in Zen. A teacher cannot teach but only aid learning. But the therapist might suggest problems like 'trying to feel a yearning for one's mother before one's own conception'. The therapist thus respects the patient's feelings and never offers interpretation of the patient's fantasies; distress must be taken at its face value. He is fully aware that these ideas taken from Zen are quite similar to the concepts used in cognitive therapy in the West. It is even possible to fit this type of therapy into a 15-minute session; moreover, such short sessions are both economical and serve to prevent the patient from developing a too intense relationship with the therapist.

Professor Kitamura at the National Institute of Mental Health in Tokyo was one of several psychiatrists I spoke to who had spent some time in either the UK or USA. In his experience there was no real difference in depression in Japan compared to England – the diagnosis was essentially similar. However, in Japan over 90 per cent of those who are depressed do not seek help, probably because of the associated stigma. He thought that in Japan general practitioners were not good at recognising depressive disorders, compared, for example, to schizophrenia. There was, he says, a reduction in depression and suicide rates during the war; people's minds were more occupied with its effects. However there was a peak following the end of the war perhaps related to the realisation that the Emperor was no longer a God.

Rates for depression are significantly lower in the East – Taiwan, for example – than in the West. Professor Chen in Hong Kong thinks the explanation is that the Chinese with their Confucian philosophy are more accepting of their fate. The Chinese family, too, is a much more supportive structure than in the West and it is quite common for a married son to live at home with his parents. Dependence on the family is not seen as a weakness. In fact, one cause of depression is separation from the family and the treatment involves bringing the patient back into the family circle. Stressful life events are thus best treated on a here-and-now basis. Antidepressant drugs like Prozac are only used for what he called atypical depression with a strong impulsive component, and electro-convulsive therapy for suicidal patients. While having low rates for depression, the Chinese had higher rates for anxiety, and this may reflect the stress imposed by family life; for while being within the family can lead to security, once outside the family anxiety could be concomitantly higher. It is also the case that tensions exist within families and these too can generate anxiety. The best way, nevertheless, to prevent depression is, in Chen's view, to stick to Chinese values.

Chinese mothers suffer from postnatal depression. As one mother put it, 'I could develop no affection for my second daughter. I was like a piece of log lost in one vast rough ocean, struggling to keep afloat.' But only about 4 per cent of mothers in Hong Kong, it is claimed, suffer from postnatal depression, compared to around 10 per cent in the West. The lower rate may reflect the Chinese practice of providing the mother with abundant social support. There is the practice of 'doing the month' after a child has been born. This involves family and friends coming into the home so that the mother, for one month, has no household chores.

Entry to the Beijing mental hospital, a little way out of town, where there is an Institute of Psychology, is by referral from doctors. There is also an outpatient clinic downtown. I was impressed by the facilities for the patients at the hospital, such as well-stocked libraries, an exercise room with modern equipment that any health club would be proud of, and a room with mirrors, where ballroom dancing is taught to the patients as part of their treatment.

Chinese patients with depression did not fit in with the diagnosis outlined in the standard Western textbooks; rather most had bodily complaints, such as backache, headache and no appetite. Less than one fifth of all those suffering from depression were ever treated. The local psychiatrists confirmed medical anthropologists' ideas on somatisation and that the expression of personal distress was in terms of bodily pain. It was suggested that this might be due to the Chinese reluctance to talk about feelings; it is actually considered to be impolite, and it is much more acceptable to complain of headache and fatigue. Chinese society is thus not one that willingly acknowledges emotional problems. One never asks a colleague how he or she feels.

Most intriguing was Morita therapy, which originated in Japan and is mainly used for patients with obsessional and anxiety disorders, but also for depression. Morita therapy does not involve teaching but learning by the patient about the nature of their illness. It is a very strict regime lasting one to two months and involves four stages. During the initial stage patients must stay in bed for about ten days. They are not allowed to talk to anyone or watch TV. They can of course collect their meals and use the toilet facilities. The doctor does not talk to the patient but may write notes. All that the patients are required to do is to spend one hour a day writing about their feelings. During the second phase they are allowed out of bed and do light labour in the yard or garden, and while some talking is permitted they must not discuss their illness. Then they move into the third phase which involves heavy labour for a further week or two. And then there is a rehabilitation stage in which they begin to integrate back into their normal workplace. This therapy is regarded as one which 'follows nature' and is thought to be very successful.

There was considerable enthusiasm for acupuncture which may be carried out daily for 30-minute sessions. There was a certain amount of irritation in the American refusal to take this Chinese-type treatment more seriously, particularly as it avoids the side effects of antidepressant drugs. One psychiatrist saw depression in terms of an imbalance of neurotransmitters which could be modified by acupuncture. Buddhism is not regarded as being much help, but it was suggested that Taoism could provide a means to prevent depression. Taoism could help one to see the world differently and adopt a dialectical stance. 'If a cup of hot tea burns your mouth, whom do you blame, yourself or the tea? The answer, of course, is to wait for it to cool.' The middle way is always regarded as the best.

It was in China that I was introduced to the idea of rational suicide, particularly by those whose lives had been destroyed by the Cultural Revolution. Their shame and guilt, and the destruction of their lives, made suicide a rational choice and did not reflect a depressive illness. And today a major problem in rural areas of China is suicide by young women. It has been suggested that suicide for these young women, who lead a very hard life, has almost become a preferred option; almost a rational choice, albeit too often an impulsive one.

My journey increased my conviction that depression is almost always related to sadness due to loss of some sort or another – money, personal relationships, social status, job, security – but that who gets depressed may be influenced by a genetic predisposition. How the depression is expressed, interpreted and treated can be very significantly affected by the prevailing culture. It may well be that, for example, there is much more somatisation of depression in non-Western cultures but also this may in fact be much more prevalent in the West than is generally recognised. It may be that there are lessons to be learned from the treatment of depression in other cultures. The claimed reduction of postnatal depression in China by 'doing the month' is a striking example. Another is the manner in which thinking in a Zen-like way may promote an especially good relationship between doctor and patient, and so short sessions of cognitive type therapy may be effective. But the only way to find out is to do reliable research, for anecdotal evidence, while it might be interesting, even suggestive, is never sufficient on its own.

CHAPTER FOURTEEN

The Future

This book, like Burton's *Anatomy of Melancholy*, has in part been a personal quest to try and understand the nature of depression, and how it can be treated and prevented. There is a great deal of information available but it would be misleading to say that depression is understood. Vulnerability to depression that can be triggered by life events is determined both by genes and early experience. Considering how important genetics is, it is disappointing how little one can say about it at this stage, other than to keep remembering its influence. As regards early experiences and life events these can be interpreted in terms of attachment theory, loss and distortions of thinking. But there is no coherent psychological theory which fully explains the nature of the depressive state. The idea of malignant sadness may, I hope, put depression in an evolutionary context and emphasise its pathological and biological nature. It also emphasises the vicious feedback loop between emotion and cognition.

It is not even very clear to me what an explanation of depression will be like. If the Good Fairy Godmother of Science allowed me a single question about the nature of depression, what would be the most important question to ask? Would it be the link between psychological processes and processes occurring in the brain? Progress in this area has been very encouraging, though there is a very long way to go. At least some of the areas of the brain involved in depression have been identified and the interactions between the amygdala and other regions of the brain are a crucial area for future research. It is now clear that, in animals at least, early experiences can alter brain processes and have long-term effects. Neurotransmitter and hormonal levels are clearly linked to depression but these are not necessarily the true causes but key links in the chain. By analogy, smoking can cause lung cancer but the cellular processes that are initiated are complex and like life events give almost no clue as to the nature of the chain of malignant processes that they initiate.

It is nevertheless in relation to chemical processes occurring in the brain that future research is very promising, together with the development of new drugs to restore normal function of the neurons involved. A very important advance would be made if tests were available to reliably diagnose and distinguish different depressive states. The criteria as laid down in DSM-IV, for example, helpful as they are, need to be improved so that there is more consistency in diagnosis by different doctors. It would also allow the progress of the illness to be properly monitored. It would be invaluable to have a reliable means for choosing the appropriate treatment; for example, which antidepressant of the many available should be prescribed.

It is thus in relation to genetics that the future holds much promise. Progress in the understanding of cancer relied very heavily on genetics and depression is an even more difficult area as several genes, perhaps many, are involved in predisposing individuals to depression, and so they will be hard to identify. Only when these genes are identified will some of the fundamental features of depression be revealed, though the revelation might not be simple as it will, in turn, probably require a proper understanding of brain function and its relation to psychological processes. But there is no other way. And when the genes are identified it will become possible to develop animal models for depression by altering the appropriate genes, and so greatly facilitate the development of new drugs. I put great emphasis on the development of new drugs as it is hard to see any major advances being made in the field of psychotherapy, particularly in view of the importance of the therapeutic alliance. But one should bear in mind the poet Paul Valery's remark that we enter the future backwards and I may be unduly pessimistic about the difficulties.

There may be some anxiety that if the genes for depression are identified then there would be serious ethical issues in relation to genetic testing. For example, it might become possible to test an individual for vulnerability to depression. I believe these fears are exaggerated, and a recent study suggests that patients with depressive disorders are generally positive about the possibilities of such tests. A large majority of manic-depressive patients said that they would take advantage of such tests; they felt that a major benefit would be that they would then be able to seek the appropriate treatment.

The question that I long to put to the Good Fairy Godmother of Science is, how would it be possible to reduce the incidence of depression? Any means that could prevent depression would be an enormous advance, for apart from the tremendous personal suffering that depression causes, the financial burden is enormous – not just the direct costs of treatment but working days lost – and runs into billions of pounds. It is crucial to recognise that depressed patients, even those with less severe depression, have more disability than persons with chronic physical diseases like heart disease, diabetes and arthritis. In the United States in 1990 the total annual cost of depression was estimated at $44 billion, of which one quarter was due to treatment and the rest to absenteeism and premature death. Drug treatment accounted for only about 10 per cent of the total treatment costs.

It has long been known that depression may worsen the disability caused by physical illness. Patients with depression have a shorter life expectancy even when suicide is taken out of the equation. For patients with a severe illness like a heart attack or stroke, the increased risk of dying if they are depressed is as high as fivefold. The reasons for this are not known but it could be that in non-depressed patients either denial or a fighting spirit are beneficial. In a follow-up of patients who had been hospitalised with depression 16 years earlier, it was found that their mortality was more than twice that of the normal population, that less than one fifth had remained well, and that over one third had an unnatural death, severe distress or handicap. There was also a high rate of readmission to hospital, nearly one half.

From my own experience I can confirm that there is a considerable stigma still attached to depression. Because I have made my own experience of depression public, many, many people have thanked me and congratulated me on my bravery. What they are really saying is that they are surprised and impressed that I am prepared to admit having an illness that is so highly stigmatised. In fact, for me, talking in public about my depression requires no bravery whatsoever as it could not have affected, and has not, my professional standing. I also have experience with students at my university. They are prepared to talk to me because they know that I have been through a similar experience and so will understand them, but they are very unwilling to let anyone else know their condition. Even so, I must admit that I am not free of the stigma, for I prefer a biological explanation for my depression rather than a psychological one.

It is probably true that the illness is poorly understood by the public at large. What in fact do the public believe about the nature and causes of depression? To what extent are lay theories similar to those of, for example, clinical psychologists? It is important to know, for the beliefs a depressed person has about the cause of their depression could influence their choice of, and response to, treatment. There is evidence that, for example, the extent to which patients in cognitive behavioural therapy believe in a cognitive model of depression and are able to relate to it, influences the rapidity of response and long-term outcome.

Lay beliefs in the causes of depression give high priority to social deprivation, interpersonal difficulties, traumatic experiences and negative self-image. Like depressed patients, lay persons have quite extensive beliefs about the causes of depression and there are few differences in the beliefs of those who are depressed and those who are not. However, depressed patients tended to show a greater preference for biological explanations than psychological ones, possibly because this made them less responsible for the condition and could justify the use of antidepressant drugs. Depressed patients do not spontaneously link distressing childhood experiences with depression. But in spite of their belief in a biological basis less than one fifth of those questioned in a large survey thought depression a condition that should be treated with drugs and most thought, wrongly, antidepressant drugs were addictive. Over 80 per cent thought counselling to be the most effective treatment. Part of the reason for the stigma attached to depression is the view that it means the person is unbalanced, or at best neurotic. For reasons like this many people who might be depressed were hesitant to consult their doctor on the grounds that the doctor would be either irritated or annoyed. They are embarrassed to seek medical help. About half of those with a depression thus do not, in fact, consult their doctor.

The stigma associated with depression, and the sense of shame and guilt in the patients that they are depressed, are still strong. In relation to postnatal depression, for example, many women feel it is a shameful and humiliating experience and want to forget it; they feel guilty for having had a depression for which they feel they were responsible. Again there are many reports, even from doctors, about the lack of support that they received from their colleagues when they suffered from either depression or manic depression.

It is thus not a surprise, though a cause for concern, that it is widely accepted in many countries that far too many people suffering from depression do not get medical help; as many as 50 per cent have been reported in the USA. Of the patients attending a GP's surgery, as many as 30 per cent are suffering from a depressive disorder. For this reason the Royal College of Psychiatrists launched a five-year Defeat Depression Campaign. Their aim was both to change public attitudes towards depression and also improve doctors' diagnoses and treatment of the condition. One of the reasons for the failure of doctors to diagnose depression may be that it is neglected in the training of doctors and so their diagnostic skills are inadequate. Another problem is that there is a tendency for GPs to prescribe too low a dose of antidepressants and not continue treatment for long enough. Older patients are often frightened of becoming addicted.

In the UK about 90 per cent of treatment of depression is by general practitioners. The vast majority of patients with mental disorders who go to their general practitioner are suffering from depression, anxiety, or a mixture of the two states. But only a minority of these patients will talk with open feelings of distress, but rather will complain of all sorts of bodily symptoms. They somatise their depression and this makes diagnosis difficult for their GP. Each GP in the UK sees about 500 depressed or anxious patients a year; that is over a million such patients a year seek help. It is a major problem. Only a very small proportion – those most severely depressed – end up in hospital, where they occupy just under a half of psychiatric beds. Probably only about half of those suffering from depression are diagnosed by their doctor and of these less than 10 per cent will be referred to a psychiatrist for specialist help. Thus most of the treatment of depressed patients is carried out by general practitioners. This is important, for the mere recognition by their doctor that they are depressed can improve the outcome. However, brief consultations in busy surgeries, together with the unwillingness of some patients to accept that they have a psychiatric problem, increase the difficulties. The failure by doctors to recognise depression may involve collusion with the patient, for it serves both their interests; it saves the patient from facing up to their problems and the shame and stigma of being depressed, and it saves the doctor the effort of taking a long and detailed history and the difficulty of facing up to dealing with the patient's complex psychosocial problems. And it also saves the doctor a most valuable and scarce resource – time.

It is a puzzle as to why psychotherapy is not always the first treatment to be recommended by doctors – except in cases where patients are so depressed that they are unable to respond to psychotherapy. Numerous trials fail in most cases to find any significant difference between treating patients with psychotherapy or an antidepressant. Given that psychotherapy is as effective in most cases as medication, and since the latter does not have serious side effects, why are not most patients given psychotherapy as the treatment of choice? This probably reflects the ease with which an antidepressant can be prescribed and taken, compared to the difficulty in finding psychotherapists, who are in short supply, and in attending the sessions. It may also reflect the doctor's lack of knowledge about the effectiveness of psychotherapy, and the time required for psychotherapy to work – patients are not always patient. The funders of health-care treatments are always looking for cost-effective procedures, but the final difference in cost is probably not significant. A major problem in the United States is the unwillingness of insurance companies to pay for psychotherapy.

The doctor is confronted with a difficult problem when the patient's condition is related to social conditions and difficult personal relationships. Should antidepressant medication be offered, and if not how can the doctor initiate social intervention? Should a doctor advise a patient to, for example, break a disabling relationship or recommend psychotherapy, if this is available? One has to keep recalling that around 40 per cent of all patients will improve with no or minimal treatment. Nevertheless, recognising that poverty, poor housing and unemployment are so common among depressed patients, help by a social worker can be beneficial. The role of the day-care hospital can also be very supportive.

There are many charitable organisations which help with depression. For example, the Depression Alliance in the UK has around 4,000 members and receives some 50,000 requests for information each year. The helpline for SANE receives about 1,000 calls a week, over 80 per cent of whom are suffering from depression. Together, they and many other organisations provide an invaluable information and support service, as well as sets of booklets on various aspects of depression which can help both those who are sufferers and those who care for them. These organisations believe that educating the public can assist both prevention and treatment. There are also of course numerous books about self-help for depression available to the public.

In relation to postnatal depression, many women find it easier to seek help from non-professionals, such as self-help groups. This puts a depressed woman in contact with someone who has had a similar experience. Women prefer not to join groups run by a professional like a psychologist. However, midwives are becoming particularly helpful in preventing postnatal depression due to the continuity of care in providing social support. Such help is very important as boys of mothers depressed in the first year following birth were tested when they were three years old and were found to have a significantly lowered intellectual attainment. Early experience of insensitive maternal interactions predicted the persistence of poorer cognitive function.

There are many attempts to treat depression which have not yet been tested in trials but which may have helped patients and these include exercise, yoga, meditation, prayer and support groups. One patient said, 'being part of a collective voice is in itself a healing thing as it is opposing prejudices and stigma, and gaining autonomy . . . by being involved I can feel the creation of a movement that will empower and heal others and thus myself'. There is even the possibility that increasing the amount of tryptophan, a molecule from which serotonin is synthesised, in the diet might help those who are vulnerable. Foods with a high tryptophan content include milk, turkey, soya beans, cottage cheese, pumpkin seeds, tofu and almonds.

Recurrent depression in an individual places a severe burden on the family. About one quarter of marriages break up if one of the partners is depressed. How can the families be helped? Initially it was thought that the family could primarily help by making sure that the patient took the antidepressant drugs. But it is now recognised that it is also necessary to help them accept the pattern of the illness and to recognise the factors that could precipitate a relapse. Ignorance about depression is a major obstacle to helping, for nothing could be more misguided than the belief that a depressed person can be helped out of their depression by distraction, or the injunction to try harder. Criticism by a carer of a depressed person can make things much worse and even cause a relapse and this knowledge puts even a greater strain on carers. Health professionals, family members or partners can play a major role in helping a depressive avoid relapse by keeping a careful watch for the symptoms that precede a relapse such as early waking and increased 'nervousness'.

A distressing number of suicides occur among patients who are already seeing a health professional. The risk of a successful suicide in the year following an unsuccessful attempt is 100 times that in the general population. There need to be improved ways of preventing suicide and recognising the signs. Organisations like the Samaritans and SANE provide some support and receive thousands of calls each year related to potential suicide. Mental health literacy is knowledge and beliefs about mental disorders and their recognition and treatment. Too many of the public cannot correctly recognise mental disorders, and many do not appreciate the physical symptoms of depression. Biological factors are underestimated compared to environmental and social ones. Even the Nuffield Council on Bioethics argue that the patient should be viewed as a whole and thus genetic influences should not be separated out from the influence of the environment. Self-help is seen by many as the most useful way forward, though exercise and cognitive therapy get some support. Medication is viewed very negatively and antidepressants are seen as having serious side effects and also leading to dependence. Natural remedies like vitamins are much more favoured. All this can lead to a failure to seek medical help. Overall, about one third of all patients and even patients with current symptoms reported nondisclosure of self-perceived psychological problems to their doctor fearing the doctor would see them as neurotic or unbalanced. The public does not reliably distinguish between psychiatrists who are medical doctors and psychologists. They also do not appreciate the differences between therapies such as psychoanalysis and cognitive therapy. The media do not help, and mental illness and psychiatrists all too often presented negatively.

Public education about depression might provide an important way forwards, but the only way to find out is to try it and see. It could positively affect how parents treat their children and enable them to recognise the risk factors. In the USA a programme has been developed that is designed to prevent depression in children in the 10-to-13 age group who are considered to be at risk. The children at risk are identified by already having some depressive symptoms as well as conduct problems and poor performance at school. The children are taught cognitive techniques to deal with their problems and preliminary results are very encouraging. Another way forward might be to teach all children about the nature of depression, and how to recognise it, in health or related lessons at school. After all, nearly one in ten of the pupils will experience a severe depression during their lifetime. In a private school in the USA typical questions after a course on mental illness include giving an opinion in 40 words as to why the increase in suicide has increased since their parents were their age, and multiple choice questions on depression such as 'It happens to adults only'. Education along similar lines has not been found to prevent depression but it did lead to students having more positive attitudes to getting help. In the UK short educational workshops with secondary school students resulted in more positive attitudes towards people with mental health problems.

What advice would I give to someone who is depressed or who is a partner of a depressive? Recognise that it is a serious illness about which one should not feel ashamed, and get help from a professional with proven experience in dealing with depression. The carer or partner should not try to minimise the problem by, for example, trying to get the sufferer to deal with the problem themselves. Try to get psychotherapy as the preferred treatment; cognitive therapy would be a sensible choice, but if that does not work, or the depression is too severe, then take antidepressant drugs until therapy becomes possible. Also, learn as much as you can about the condition so that you can ask whoever is providing treatment the right questions. Ultimately, those who are depressed have to take responsibility for their own condition and so need to understand as much about it as possible. And if possible, take the advice given at the end of Burton's *Anatomy of Melancholy*: 'Be not idle'.

References

Most topics are well covered by the books listed under the General heading. More specialised topics and recent results are listed under chapter headings.

GENERAL

Checkley, S. (ed.) (1998), *The management of depression*. Blackwell Science, Oxford.

Hammer, C. (1997), *Depression*. Psychology Press, Hove.

Honig, A., and van Praag, H. M. (eds.) (1997), *Depression: neurobiological, psychopathological and therapeutic approaches*. Wiley, Chichester.

Paykel, E. S. (ed.) (1992), *Handbook of affective disorders*. Churchill Livingstone, Edinburgh.

INTRODUCTION

Burton, R. (1651), *The Anatomy of melancholy*. Ed. T. C. Faulkner et al. 1989–94. Clarendon Press, Oxford.

CHAPTER 1: The Experience of Depression

Gotlib, I. H. and Hammen, C. I. (eds.) (2002), *Handbook of depression*. Guildford Press, New York.

Jackson, S. W. (1986), *Melancholia and depression: from Hippocratic times to modern times*. Yale University, New Haven.

Porter, R. (ed.) (1991), *The Faber book of madness*. Faber, London.

Solomon, A. (1998), 'Anatomy of melancholy.' *The New Yorker*, 12 January, 46–61.

Styron, W. (1991), *Darkness visible*. Picador, London.

Wong, M.-L. and Licinio, J. (2001), 'Research and approaches to depression.' *Nature Reviews: Neuroscience* 2, 343–351.

CHAPTER 2: Defining and Diagnosing Depression

Diagnostic and Statistical Manual of Mental Disorders (DSM-IV), 4th edn., American Psychiatric Association, Washington.

Goodyer, I. (1993), 'Depression among pupils at school.' *British Journal of Special Education* 20, 51–4.

CHAPTER 3: Mania

Jamison, K. R. (1995), *An unquiet mind*. Picador, London.

Winokour, G. (1991), *Mania and depression. A classification of syndrome and disease.* John Hopkins University Press, Baltimore.

CHAPTER 4: Other Cultures

Jadhav, S. (1996), 'The cultural origins of Western depression.' *International Journal of Social Psychiatry*, 42, 269–86.

Kleinman, A. (1988), *Rethinking psychiatry: from cultural category to personal experience*. Free Press, New York.

Kleinman, A., and Cohen, A. (1997), 'Psychiatry's global challenge.' *Scientific American*, May, 74–7.

Kleinman, A., and Good, B. (eds.) (1985), *Culture and depression.* University of California Press, Berkeley.

Littlewood, R., and Lipsedge, M. (1997), *Aliens and Alienists: ethnic minorities and psychiatry*. Unwin Hyman, London.

Raguram, R. et al. (1996), 'Stigma, depression and somatization in South India.' *American Journal of Psychiatry* 153, 1043–9.

Weiss, M. G. et al. (1995), 'Cultural dimensions of psychiatric diagnosis.' *British Journal of Psychiatry* 166, 353–9.

CHAPTER 5: Who Gets Depressed and Why?

Bland, R. C. (1997), 'Epidemiology of affective disorders.' *Canadian Journal of Psychiatry* 42, 367–77.

Caspi, et al. (2003), 'Influence of life stress on depression: moderation by a polymorphism in the 5-HTT gene.' *Science* 301, 386–389.

Cooper, P. J., and Murray, L., 'Postnatal depression.' *British Medical Journal* 316, 1884–6.

Faraone, S. U., and Biederman, J. (1998), 'Depression: a family affair.' *Lancet* 351, 158.

Fombonne, E. (1994), 'Increased rates of depression: update of epidemiological findings and analytical problems.' *Acta Psychiatrica Scandinavica* 1994, 1–12.

Harris, T., and Brown, G. W. (1996), 'Social causes of depression.' *Current Opinions in Psychiatry* 9, 3–10.

Katona, C. et al. (2005), 'Pain symptoms in depression: definition and clinical significance.' *Clinical Medicine* 5, 390–395.

Kessler, R. C. et al. (1997), 'Prevalence, correlates, and course of minor depression and major depression in the national comorbidity survey.' *Journal of Affective Disorders* 45, 19–30.

Kessler, R. C., and Magee, W. J. (1993), 'Childhood adversities and adult depression: basic patterns of association in a US national survey.' *Psychological Medicine* 23, 679–90.

Kleinman, A. (1996), 'China: The Epidemiology of Mental Illness.' *British Journal of Psychiatry* 169, 129–30.

Lepin, J.-P. et al. (1997), 'Depression in the community: the first pan-European study.' *International Clinical Psychopharmacology* 12, 19–20.

McGuffin, P. et al. (1996), 'A hospital-based twin register of the heritability of DSM-IV unipolar depression.' *Archives General Psychiatry* 53, 129–36.

Paykel, E. S. (1991), 'Depression in women.' *British Journal of Psychiatry* 158, (suppl.) 22–9.

Robertson, M. M., and Katona, C. L. E. (eds.) (1997), *Depression and Physical Illness*. Wiley, Chichester.

Weissman, M. M., and Olfson, M. (1995), 'Depression in women: implications for health care research.' *Science* 269, 799–801.

Wittchen, H.-A. et al. (1994), 'Lifetime risk of depression.' *British Journal of Psychiatry* 165 (suppl. 26), 16–22.

Wurtman, R. J., and Wurtman, J. J. (1989), 'Carbohydrates and depression.' *Scientific American*, January, 50–7.

CHAPTER 6: Suicide

Alvarez, A. (1971), *The Savage God. A study of suicide*. Penguin, London.

Brown, P. (1997), 'No way out.' *New Scientist*, 22 March, 34–7.

Holden, C. (1992), 'New discipline probes suicide's multiple causes.' *Science* 256, 1761–2.

Jamison, K. R. (1999), *Night falls fast: Understanding suicide*. Picador, London.

Takahashi, Y. (1997), 'Culture and suicide: from a Japanese psychiatrist's perspective.' *Suicide and Life Threatening Behaviour* 27, 137–45.

CHAPTER 7: Emotion, Evolution and Malignant Sadness

Izard, C. E. (1991), *The psychology of emotions*. Plenum, London.

Jamison, K. R. (1994), *Touched with fire. Manic depressive illness and the artistic temperament*. Free Press, New York.

Keller, M. C., and Nesse, R. M. (2005), 'Is low mood an adaptation? Evidence for subtypes with symptoms that match precipitants.' *Journal of Affective Disorders* 86, 27–35.

Nesse, R. M., and Williams, G. C. (1995), *Evolution and healing*. Weidenfeld and Nicolson, London.
Pinker, S. (1997), *How the mind works*. Allen Lane. The Penguin Press, London.
Price, J. et al. (1994), 'The social competition hypothesis of depression.' *British Journal of Psychiatry* 164, 309–15.
Watson, P., and Andrews, P. (2002), 'Toward a revised evolutionary adaptationist analysis of depression: the social navigation hypothesis.' *Journal of Affective Disorders* 72, 1–14.

CHAPTER 8: Psychological Explanations

Beck, A. (1991), 'Cognitive therapy: a 30-year retrospective.' *American Psychology* 46, 368–75.
Bowlby, J. (1981), *Attachment and loss*. Volume III: *Loss: sadness and depression*. Penguin, London.
Brenner, C. (1991), 'A psychoanalytic perspective on depression.' *Journal of the American Psychoanalytic Association* 39, 25–42.
Brewin, C. R. (1996), 'Theoretical foundation of cognitive behaviour therapy for anxiety and depression.' *Annual Review of Psychology* 47, 33–57.
Brewin, C. R. (1996), 'Cognitive processing of adverse experiences.' *International Review of Psychiatry* 8, 333–9.
Ekman, P. (1992), 'An argument for basic emotion.' *Cognition and Emotion* 6, 109–200.
Ellis, H. D. et al. (1996), 'Delusional misidentification of inanimate objects.' Cognitive Neuropsychiatry 1, 27–40.
Goldman, D. (1996), 'High anxiety.' *Science* 274, 1483.
Holmes, J. (1993), *John Bowlby and Attachment Theory*. Routledge, London.
Kristeva, J. (1989), *Black Sun*. Columbia University Press.
Parkes, C. M. (1996), *Bereavement. Studies of grief in adult life*. Routledge, London.
Parkes, C. M. et al. (1996), *Cross-cultural studies of death and bereavement*. Routledge, London.
Peterson, C. et al. (1993), *Learned helplessness*. Oxford University Press, Oxford.
Rutter, M. (1995), 'Clinical implications of attachment concepts: retrospect and prospect.' *Journal of Child Psychology and Psychiatry* 36, 549–71.
Sanders, C. M. (1989), *Grief: The Mourning After*. Wiley, New York.
Steele, H., and Steele, M. (1998), 'Attachment and psychoanalysis: time for a reunion.' *Social Development* 7, 92–118.
Suomi, S. J. (1997), 'Early determinants of behaviour: evidence from primate studies.' *British Medical Bulletin* 53, 170–84.

Williams, J. M. G. (1992), 'Autobiographical memory and emotional disorders.' In S. A. Christianson (ed.), *Handbook of Emotion and Memory*. Erblaum, New Jersey.

CHAPTER 9: Biological Explanations and the Brain

Damasio, A. R. (1997), 'Towards a neuropathology of emotion and mood.' *Nature* 386, 769–70.

Drevets, W. C. (1998), 'Functional neuro imaging studies of depression.' *Annual Review of Medicine* 49, 341–61.

Fink, G. et al. (1998), 'Sex, hormones, your mood and memory.' *Clinical and Experimental Pharmacology and Physiology* 25, 764–75.

Goodyer, I. M. et al. (1998), 'Adrenal steroid secretion and major depression in 8-to-16-year-olds.' *Psychological Medicine* 28, 265–73.

Herbert, J. (1997), 'Stress, the brain and mental illness.' *British Medical Journal* 318, 530–5.

Hyman, S. E. (1998), 'A new image for fear and emotion.' *Nature* 393, 417–18.

Kramer, M. S. et al. (1998), 'Distinct mechanism for antidepressant activity by blockade of central substance P receptors.' *Science* 281, 1640–5.

Knutson, B. et al. (1998), 'Selective alteration of personality and social behaviour by serotonergic intervention.' *American Journal of Psychiatry* 15, 373–9.

Le Doux, J. (1998), *The emotional brain*. Weidenfeld and Nicolson, London.

Mlot, C. (1998), 'Probing the biology of emotion.' *Science* 280, 1005–7.

Nemeroff, C. B. (1998), 'The neurobiology of depression.' *Scientific American*, June 1998, 28–35.

Öngür, D. et al. (1998), 'Glial reduction in the sub-genual prefrontal cortex in mood disorders.' *Proceedings of the National Academy of Science* 95, 13290–5.

Sapolsky, R. M. (1997), 'The importance of the well-groomed child.' *Science* 277, 1620–1.

Schiepers, O. J., Wichers, M. C., and Maes, M. (2005), 'Cytokines and major depression.' *Prog Neuropsychopharmacol Biol Psychiatry* 29, 201–17.

CHAPTER 10: Antidepressants and Physical Treatments

Briley, M., and Montgomery, S. (eds.) (1998), *Antidepressant therapy at the dawn of the third millennium*. Martin Dunitz, London.

Brown, G. K. et al. (2005), 'Cognitive therapy for the prevention of suicide attempts: a randomized controlled trial.' *JAMA* 294, 563–70.

Cipriani, A., Barbui, B., and Geddes, J. R. (2005), 'Suicide, depression and antidepressants.' *British Medical Journal* 330, 373–374.

Frank, L. R. (1978), *The history of shock treatment*. Frank Leroy: San Francisco.
Freeman, C. P. (ed.) (1995), *The ECT handbook*. Royal College of Psychiatrists, London.
Healy, D. (1997), *The antidepressant era*. Harvard University Press, Cambridge.
Mayberg, H. et al. (2005), 'Deep brain stimulation for treatment-resistant depression.' *Neuron* 45, 651–60.
Thase, M. E. et al. (2005), 'Remission rates following antidepressant therapy with bupropion or selective serotonin reuptake inhibitors: a meta-analysis of original data from 7 randomized controlled trials.' *Journal of Clinical Psychiatry* 66, 6974–81.
Wurtzel, E. (1994), *Prozac Nation*. Riverhead, New York.

CHAPTER 11: Psychotherapy

Blackburn, I. D. M, and Davidson, K. (1995), *Cognitive therapy for depression and anxiety*. Blackwell Science, Oxford.
Paykel, E. S. (1994), 'Psychological therapies.' *Acta Psychiatrica Scandinavica* 89, (suppl.) 383, 35–41.
Shapiro, D. (1995), 'Finding out how psychotherapists help people change.' *Psychotherapy Results* 5, 1–21.

CHAPTER 12: What Works?

Antonuccio, D. et. al. (1997), 'Depression: psychotherapy is the best medicine.' *The Therapist* 4, 30–40.
Brown, W. A. (1998), 'The placebo effect.' *Scientific American*, January, 68–73.
Geddes, J. R. (1999), 'Depressive disorders in adults.' *Clinical Evidence*, I, 45–55, BMJ Publishing Group, London.
Horgan, J. (1996), 'Why Freud isn't dead.' *Scientific American*, December, 74–9.
Knowing Our Minds. A survey of how people in emotional distress take control of their lives. (1997), Mental Health Foundation, London.
Roth, A., and Fonagy, P. (1996), *What works and for whom? A critical review of psychotherapy research*. Guildford, New York.

CHAPTER 13: An Excursion to the East

Chen, C.-E. (1995), 'Anxiety and depression: east and west.' *International Medical Journal* 3, 3–5.
Ono, Y., and Berger, D. (1995), 'Zen and the art of psychotherapy.' *Journal of Practical Psychology and Behavioural Health*, November, 203–10.

CHAPTER 14: The Future

Eisenberg, L. (1992), 'Treating depression and anxiety in primary care. Closing the gap between knowledge and practice.' *New England Journal of Medicine* 326, 1080–3.
Jaycox, L. H. et al. (1994), 'Prevention of depressive symptoms in school children.' *Behav. Res. Ther.* 32, 801–16.
Jorm, A. F. (2000), 'Mental health literacy.' *British Journal of Psychiatry* 177, 396–401.
Kuyken, W. et al. (1992), 'Causal beliefs about depression in depressed patients, clinical psychologists and lay persons.' *British Journal of Medical Psychology* 65, 257–68.
Paykel, E. S. et al. (1997), 'The Defeat Depression Campaign: psychiatry in the public arena.' *American Journal of Psychiatry* 154, 59–65.
Pinfold, V. et al. (2003), 'Reducing psychiatric stigma and discrimination: evaluation of educational interventions in UK secondary schools.' *British Journal of Psychiatry* 182, 342–346.
'Stigma of Mental Illness.' (1999), *Lancet* 352, 1048–1059.
Trippitelli, C. L. (1998), 'Pilot study on patients' and spouses' attitudes towards potential genetic testing for bipolar disorders.' *American Journal of Psychiatry* 155, 899–904.

Index